THE IMPOSSIBLE ART OF GOLF

THE
IMPOSSIBLE
ART OF
GOLF

An Anthology of Golf Writing

Selected by
ALEC MORRISON

Oxford   New York
OXFORD UNIVERSITY PRESS
1994

Oxford University Press, Walton Street, Oxford OX2 6DP

Oxford New York
Athens Auckland Bangkok Bombay
Calcutta Cape Town Dar es Salaam Delhi
Florence Hong Kong Istanbul Karachi
Kuala Lumpur Madras Madrid Melbourne
Mexico City Nairobi Paris Singapore
Taipei Tokyo Toronto
and associated companies in
Berlin Ibadan

Oxford is a trade mark of Oxford University Press

British Library Cataloguing in Publication Data
Data available

Library of Congress Cataloging in Publication Data
The Impossible art of golf : an anthology of golf writing /
selected by Alec Morrison.
p. cm.
Includes bibliographical references.
1. Golf. 2. Golf stories.
I. Morrison, Alec.
GV965.I46 1994 796.352—dc20 94–8831
ISBN 0–19–211698–3

1 3 5 7 9 10 8 6 4 2

Typeset by Graphicraft Typesetter Ltd., Hong Kong
Printed in Great Britain
on acid-free paper by
The Bath Press
Bath, Avon

CONTENTS

INTRODUCTION

❖

A certain wariness seems the appropriate tone for an introduction to any anthology. At least it is apparent to some degree in all of those that I have read. The inescapable twin prongs of error, the inclusion of too many pieces deemed too well known by some and the omission of too many which are cherished as old favourites by others, ensures a critical reception. In my own case, the wariness is even more appropriate. Though I have often sat in the same chair in which Bernard Darwin is said to have composed many of his articles, I can make no claim that the nature of the contact has improved either my swing or my literary style. In my defence I can only truly plead an enthusiasm for the game and its literature, which has lasted throughout my grown-up life and which may be unusual in the equality of its emphasis.

Golf shares with cricket and baseball the distinction of having an enormously rich literary heritage. The volume of writing about the game is huge. Much of it is instructional, since more than with any other sport its techniques are subject to minute analysis. Book after book prescribes infallible solutions to the problems of every shot, from drive to putt. If only we could put them into practice, or even if they could agree on which portion or position of our anatomy it is in which the great secret is to be found, how happy and successful we should be. None of this material has seemed worth including here. Similarly omitted in very large part are ghosted memoirs, 'autobiographies', and other writing of this sort, though it is not always easy to establish the existence of a ghost for certain.

The reason for these omissions is that they nearly always fail the test that I have tried to apply to every item that is included—the essential quality of readability. The question that has been asked of every piece is: does it contain that spark of style or content or context that makes for delight in its discovery or rediscovery?

The bias therefore is literary. It seems particularly pleasing to me when a writer reveals an affection for the game and uses it as either a subject in itself or an episode in a longer work. Not all writers, of course, are well disposed. According to J. L. Carr, Max Beerbohm subscribed a shilling to W. G. Grace's testimonial 'not in support of cricket but as an earnest protest against golf', and Trollope's reported behaviour at St Andrews also suggests a lack of sympathy for the game or at least its etiquette. In 1906 Arthur Benson, the Master of Magdalene College, Cambridge, was already noting with dismay the tendency among undergraduates to 'call the country dull because one has not the opportunity of hitting and pursuing a little white ball round and round among the fields, with elaborately contrived obstacles to try the skill and temper'.

Looking back, it is clear to me now that there were two sources from which the main lines of this collection developed. The first of these is Thomas Peter's description of the great challenge match between the Dunn brothers and the St Andrews men in 1849. This account was contained in a typescript which presumably belonged to Bernard Darwin and which was recently found at the Dormy House Club in Rye. For me it has every necessary ingredient of the very best sports reporting —the evident excitement of the writer, the setting of the scene, the changing fortunes, the final loss of nerve by the Dunns which costs them the match. And all of it told in a style which is still fresh and true today.

The other source is Ford Madox Ford's description of Tietjens in the sand-hills at Rye. It is clear from his letters that Ford was a golfer and he obviously knew the Rye course as it then was, and clearly appreciated one of the unique pleasures of links golf, namely the sense of being alone in a landscape of rolling dunes and sea under a hemisphere of sky.

Sir Walter Simpson wrote in 1887 that there were no worthwhile golfing novels and from everything that I have read this still seems eminently true today. Bernard Darwin went further and maintained that the humorous short stories, including his own, were of no value and that they would, in most cases have been made funnier by the removal of the golfing setting. This seems unnecessarily harsh, though even Darwin exempted P. G. Wodehouse from his general condemnation.

But whatever one's feelings in this respect, there can be no doubt that writers of fiction have captured that mixture of physical and aesthetic pleasure that the game can provide in a way that no mere reporting can ever do. Updike's Rabbit, Tietjens in the sand-hills, and Walker Percy's Will Barrett in the scarlet maples of Carolina are all brothers beneath the skin to each other and to all the great company of golfers who love the game.

Around the original sources a number of others have inevitably congregated, each of them reflecting some aspect of the game and its many attractions. Which those are will be different, in degree of importance at least, for all of us and perhaps different for every time we play, but I believe that the ingredients can be identified even if the exact mix cannot. Certainly for me they are the sense of the heritage and history of the game, the maddening, fascinating difficulty of it, the combination of mind and body required to tackle it, the beauty of the setting, and the social rewards of the company of one's fellow players.

It is an extraordinary, unique fact about golf that since most courses are members' clubs and some of the finest are even public a large number of us can play over the same ground where the greatest players of past and present have performed. While few of us can hope ever to step up to the plate at Yankee Stadium or out on to the Centre Court at Wimbledon and sense the shades of Tilden or Perry or Lenglen or Billie Jean King, anyone at all can play at St Andrews or, if he saves up, at Pebble Beach and, with a little luck and planning and a friend or two, at Merion, Royal St George's, and others of the classic courses.

This possibility of literally following in the footsteps of the giants is just one of the things that keeps the heritage of the game so strong and vital. To stand on the first tee of the Old Course at St Andrews is to know that you are a tiny, transient part of a great and continuing tradition. To face the second shot to the seventeenth, the Road Hole—even if you foozle it—is to share something of the essence of a Nicklaus or a Taylor. Most poignantly of all, to realize that the small, dapper figure at the 1993 British Open is that of Gene Sarazen, who won that same tournament sixty years before just down the road at Prince's and who actually played two rounds with Vardon in the North East Open in 1923, wonderfully underlines the strength and directness of the link with what can otherwise seem a legendary past.

As to the difficulty of the game, no one who has ever tried to play it will be in any doubt about that. Winston Churchill, after what was presumably an unsuccessful round, commented that the golf-club was an instrument poorly adapted to its purpose. Sometimes we all feel that the whole business of swinging a yard-long club in order to strike a small ball accurately at right angles to the direction in which we are facing is, like the flight of the bumble-bee, clearly impossible. To be able to do it, time after time, while standing with one foot six inches higher than the other and in a cross-wind, is close to magic. Yet this is what the professionals achieve and what, therefore, the weekend player aspires to.

All of us believe that our good shots are our norm, the bad ones the aberrations. For most, of course, the opposite is the truth, but the glory of the game is that there always are good shots. Just as the pro, while apparently playing immaculately, will, if he is honest, acknowledge like Hogan that he hit only a handful of shots truly perfectly, so the weekender will always have something to remember with pleasure: a longer-than-usual drive that found the fairway, a chip from the rough that hit the green, a decent putt. Every now and again we are all allowed a nibble at the carrot of success to keep us going and to renew our zest for the next round.

As if the physical difficulties were not sufficient, they are, as every golfer knows, multiplied many times by the mental stresses. The three-foot putt which is a tap-in on the practice green can seem downright impossible towards the end of a promising medal round. The ball is not going to move, as every teaching pro has said to every beginner, but few of us can avoid swinging at it too fast when we feel a little pressure.

And yet . . . and yet there is no equivalent in any other sport to the pleasure of a well-struck shot, the smooth climb of the ball into the sky, its run along the fairway or obedient, fluttering descent on to the green. The combination of the physical sensation of the hit and the visual reward of the flight is unmatched.

The discipline of most games demands that the playing areas must be at least roughly consistent with each other. Even though the length of the boundaries may differ, every cricket-pitch is twenty-two yards long, baseball bases are all ninety feet apart, every American football field is a hundred yards long. But every golf-course is different, and every hole on every course is different, and this diversity is as fundamental to the game as is the consistency of the pitches on which the others are played. Meeting and overcoming the problems of the terrain is part of the essence of golf.

Watching the professionals, it is clear that they have all mastered the problems of striking the ball to virtually the same level of extraordinary excellence. What distinguishes the true greats is partly their ability to master themselves, to control their emotions under the severest pressure, and partly their ability to manage the course on which they are playing; to manœuvre the ball around it so as always to leave themselves the easiest possible next shot.

The best courses, then, are those in which these problems are set in the most varied and subtle ways. In the beginning, on the open links of the Scottish courses, there was no real problem; the natural humps and hollows and the inevitable, changing wind provided tests enough. Bobby Jones himself came to believe that this type of course represents the finest test of golf, and the true arena for the great occasions of the game. However, where no such natural advantages are available, the architect must fall back on the bulldozer to augment the features of the land with which he has to work, and it is fascinating to see the various ways in which this challenge has been met.

Recently there has been added to these sorts of considerations the concept of the 'penal' course with spectacularly difficult holes, like Pete Dye's tiny island green, which is linked to the tee only by a narrow causeway. No doubt such holes are a fine test of nerve, and they may in part be a legitimate response by the designer to the ever-increasing skill of the players and, even more, to their ever-improving equipment. But there is still an uneasy feeling that the marketing men have begun to take over from the architects in order to try to create an outrageous challenge that large numbers of visitors will want to come to play just once, so as to be able to glory in the telling of their success or, more often, in the number of their failed attempts.

Such extreme challenges seem to negate the other vital thread in the true nature of the game—the continuing, developing sociability of the club. It is easy in all the talk of the technical difficulties of accurate striking, the problems of controlling the mind and managing the course, to overlook the supreme fact that it is a game played in company, and that it is from that company that much of the pleasure will be derived. Ever since the earliest democratic days in Scotland, the members' club has been the backbone of the game. Today there is a feeling that the traditional club is under threat. As the popularity of the game increases, and more and more courses are set up on a purely financial basis, with a premium on members who can afford to pay the necessarily large entry fees and subscriptions, rather than on the members' kinship with each other, so there is a clear danger that the traditional spirit and courtesies which are so large a part of the pleasure for many of us are in danger of being lost. When the friendly foursome is superseded by interminable competitions, the heart of the game goes with it.

Against this background, this selection covers golf from its earliest days in Scotland, through the major changes brought about mainly by the development of the ball, to its present status as a truly international sport. It does not attempt to provide a comprehensive history of the game, nor a complete gallery of its greatest players. These are available elsewhere, and in any case must fall rapidly out of date. What it does seek to do, and what surprisingly seems not to have been done quite in this way before, is to convey the gloriously rich and varied reflection of the game as it has appeared on the page in prose and poetry, whether it has been written about for its own sake or as part of some larger work.

Inevitably, given the history of the game, the material dates mostly from after 1890 and is drawn from both British and American writers. It is arranged in nine sections; a broadly chronological order has been followed in each, but where a particular juxtaposition of items has seemed to add a special point of interest I have not hesitated to break this self-imposed rule.

All sport has been called a 'magnificent triviality'. When it is illuminated by the talent of a writer with knowledge and love of the game, the pleasure it can give is made deeper and longer-lasting. It is the celebration of that pleasure that is the main aim of this anthology.

ALEC MORRISON

THE JOY OF THE GAME

❖

Defining and celebrating the fascination of golf has exercised a multitude of writers over the years. In the 1890s, when the great growth of the game outside its native Scotland was just beginning, there was a missionary zeal about the work of authors like Sir Walter Simpson, Horace Hutchinson, and Arnold Haultain. This was particularly true in the case of Arthur Balfour, whose political eminence gave him an exceptional platform from which to publicize the glories of the game. Now, with the enormous growth in the number of players, the missionary work has been well and truly done, but the game continues to inspire its devotees to try to describe its hold over them and the delight which it gives to them. This first sentence could reasonably stand inscribed over the doorway of any golf clubhouse in the world.

A tolerable day, a tolerable green, a tolerable opponent, supply, or ought to supply, all that any reasonably constituted human being should require in the way of entertainment. With a fine sea view, and a clear course in front of him, the golfer should find no difficulty in dismissing all worries from his mind, and regarding golf, even it may be very indifferent golf, as the true and adequate end of man's existence. Care may sit behind the horseman, she never presumes to walk with the caddie. No inconvenient reminiscences of the ordinary workaday world, no intervals of weariness or monotony interrupt the pleasures of the game. And of what other recreation can this be said?   A. J. BALFOUR, *The Humours of Golf,* 1890

❖

Three things there are as unfathomable as they are fascinating to the masculine mind: metaphysics, golf and the feminine heart. The Germans, I believe, pretend to have solved some of the riddles of the first, and the French to have unravelled some of the intricacies

of the last; will someone tell us wherin lies the extraordinary fascination of golf?   ARNOLD HAULTAIN, *The Mystery of Golf*, 1910

❖

It is a game for the many. It suits all sorts and conditions of men.
The strong and the weak, the halt and the maimed, the octogen-
arian and the boy, the rich and the poor, the clergyman and the
infidel, may play every day, except Sunday. The late riser can play
comfortably, and be back for his rubber in the afternoon; the
sanguine man can measure himself against those who will beat
him; the half-crown seeker can find victims, the gambler can bet,
the man of high principle, by playing for nothing, may enjoy himself, and yet feel good. You can brag, and lose matches; depreciate
yourself, and win them. Unlike the other Scotch game of whisky-
drinking, excess in it is not injurious to the health.

Golf has some drawbacks. It is possible, by too much of it, to
destroy the mind . . . For the golfer, Nature loses her significance.
Larks, the casts of worms, the buzzing of bees, and even children
are hateful to him. I have seen a golfer very angry at getting into
a bunker by killing a bird, and rewards of as much as ten shillings
have been offered for boys maimed on the links. Rain comes to be
regarded solely in its relation to the putting greens; the daisy is
detested, botanical specimens are but 'hazards', twigs 'break clubs'.
Winds cease to be east, south, west, or north. They are ahead,
behind, or sideways, and the sky is bright or dark, according to the
state of the game.   WALTER SIMPSON, *The Art of Golf*, 1887

To some minds the great field which golf opens up for exaggeration is its chief attraction. Lying about the length of one's drives
has this advantage over most forms of falsehood, that it can scarcely be detected. Your audience may doubt your veracity, but they
cannot prove your falsity. Even when some rude person proves your
shot to be impossibly long, you are not cornered. You admit to an
exceptional loft, to a skid off a paling, or, as a last appeal to the
father of lies, you may rather think that a dog lifted your ball.
'Anyhow,' you add conclusively, 'that is where we found it when
we came up to it.'   Ibid.

The game of golf is full of consolation. The long driver who is
beaten feels that he has a soul above putting. All those who can-

not drive 30 yards suppose themselves to be good putters. Your hashy player piques himself on his power of recovery. The duffer is a duffer merely because every second shot is missed. Time or care will eliminate the misses, and then! Or perhaps there is something persistently wrong in driving, putting or approaching. He will discover the fault, and then! Golf is not one of those occupations in which you soon learn your level. There is no shape nor size of body, no awkwardness nor ungainliness, which puts good golf beyond one's reach. There are good golfers with spectacles, with one eye, with one leg, even with one arm. None but the absolutely blind need despair. It is not the youthful tyro alone who has cause to hope. Beginners in middle age have become great and, more wonderful still, after years of patient duffering, there may be a rift in the clouds. Some pet vice which has been clung to as a virtue may be abandoned, and the fifth-rate player burst upon the world as a medal winner. In golf, whilst there is life there is hope.

Ibid.

❖

THE GREAT SECRET

When he reads of the notable doings of famous golfers, the eighteen-handicap man has no envy in his heart. For by this time he has discovered the great secret of golf. Before he began to play he wondered wherein lay the fascination of it; now he knows. Golf is so popular simply because it is the best game in the world at which to be bad.

Consider what it is to be bad at cricket. You have bought a new bat, perfect in balance; a new pair of pads, white as driven snow; gloves of the very latest design. Do they let you use them? No. After one ball, in the negotiation of which neither your bat, nor your pads, nor your gloves came into play, they send you back into the pavilion to spend the rest of the afternoon listening to fatuous old stories of some old gentleman who knew Fuller Pilch. And when your side takes the field, where are you? Probably at long leg both ends, exposed to the public gaze as the worst fieldsman in London. How devastating are your emotions. Remorse, anger, mortification, fill your heart; above all, envy—envy of the lucky immortals who disport themselves on the green level of Lord's.

Consider what it is to be bad at lawn tennis. True, you are allowed to hold on to your new racket all through the game, but how often

are you allowed to employ it usefully? How often does your partner cry 'Mine!' and bundle you out of the way? You may spend
the full eighty minutes in your new boots, but your relations with
the ball will be distant. They do not give you a ball to yourself at
football.

But how different a game is golf. At golf it is the bad player who
gets the most strokes. However good his opponent, the bad player has the right to play out each hole to the end; he will get more
than his share of the game. He need have no fears that his new
driver will not be employed. He will have as many swings with it
as the scratch man; more, if he misses the ball altogether upon one
or two tees. If he buys a new niblick he is certain to get fun out
of it on the very first day.

And, above all, there is this to be said for golfing mediocrity—
the bad player can make the strokes of the good player. The poor
cricketer has perhaps never made fifty in his life; as soon as he
stands at the wickets he knows that he is not going to make fifty
today. But the eighteen-handicap man has some time or other
played every hole on the course to perfection. He has driven a ball
250 yards; he has made superb approaches; he has run down the
long putt. Any of these things may suddenly happen to him again.
And therefore it is not his fate to have to sit in the club smoking-
room after his second round and listen to the wonderful deeds
of others. He can join in too. He can say with perfect truth, 'I once
carried the ditch at the fourth with my second', or 'I remember
when I drove into the bunker guarding the eighth green', or even
'I did a three at the eleventh this afternoon'—bogey being five. But
if the bad cricketer says, 'I remember when I took a century in
forty minutes off Lockwood and Richardson', he is nothing but a
liar.

For these and other reasons golf is the best game in the world
for the bad player. And sometimes I am tempted to go further
and say that it is a better game for the bad player than for the good
player. The joy of driving a ball straight after a week of slicing, the
joy of putting a mashie shot dead, the joy of even a moderate
stroke with a brassie; best of all, the joy of the perfect cleek shot—
these things the good player will never know. Every stroke we bad
players make we make in hope. It is never so bad but it might have
been worse; it is never so bad but we are confident of doing better next time. And if the next stroke is good, what happiness fills

our soul. How eagerly we tell ourselves that in a little while all our strokes will be as good.

What does Vardon know of this? If he does a five hole in four he blames himself that he did not do it in three; if he does it in five he is miserable. He will never experience that happy surprise with which we hail our best strokes. Only his bad strokes surprise him, and then we may suppose that he is not happy. His length and accuracy are mechanical; they are not the result, as so often in our case, of some suddenly applied maxim or some suddenly discovered innovation. The only thing which can vary in his game is his putting, and putting is not golf but croquet.

But of course we, too, are going to be as good as Vardon one day. We are only postponing the day because meanwhile it is too pleasant to be bad. And it is part of the charm of being bad at golf that in a moment, in a single night, we may become good. If the bad cricketer said to a good cricketer, 'What am I doing wrong?' the only possible answer would be, 'Nothing particular, except that you can't play cricket.' But if you or I were to say to our scratch friend, 'What am I doing wrong,' he would reply at once, 'Moving the head' or 'Dropping the right knee' or 'Not getting the wrists in soon enough', and by tomorrow we should be different players. Upon such a little depends, or seems to the eighteen-handicapper to depend, excellence in golf.

And so, perfectly happy in our present badness and perfectly confident of our future goodness, we long-handicap men remain. Perhaps it would be pleasanter to be a little more certain of getting the ball safely off the first tee; perhaps at the fourteenth hole, where there is a right of way and the public encroach, we should like to feel that we have done with topping; perhaps—

Well, perhaps we might get our handicap down to fifteen this summer. But no lower; certainly no lower.

A. A. MILNE, *Not that it Matters*, 1919

Not least of the delights are, of course, those few, wonderful occasions when we are touched by the divine hand and play the kind of golf we never knew we could.

I will make the account of this masterful round brief, but only out of consideration for those lobster-tanned veterans who took up the game at the age of eight, progressed by kangaroo stages to a two handicap and have deteriorated so alarmingly with each decade that when asked to exercise total recall and recount the history of some of their triumphs they reply, like Bing Crosby, 'Total recall? I can't even recall when I last broke 80.'

Well, sir, it was a glittering Sunday afternoon in the late fall. I was sitting out on our terrace on the North Fork of Long Island facing the deep blue waters of Peconic Bay and feeling pretty hot under the collar, I can tell you. We had had three people to lunch. We were still having them to lunch and it was twenty-five minutes to four, a whole hour past the time when I normally have bussed my wife with resounding gratitude, torn off to a links course twelve miles away and started to lace long drives and cunning little pitch-es—alone, of course—into the declining sun. I had already given rather gross notice to these dawdlers of my intentions by retreat-ing to my bedroom and going through my normal medical rou-tine as a senior golfer (we shall come to that a little later on).

Our guests—my lawyer, his wife and sister—would have been, at any other time, enchanting company. In fact, until they started on their third lobster at three p.m. I was convinced that no more amiable guests had ever wolfed our vodka or darkened our towels.

Far off, across the old Colonial meadows at our back, a church clock struck three-thirty. The old Colonial churches still have their uses, not the least of which is to toll the knell of parting guests. My lawyer turned to me and said: 'A beautiful time of day, this, espe-cially in the fall.' In a flash, I saw my opportunity for a cliffhanger's cry. 'Yes, indeed,' I said, 'you know, a friend of mine now patter-ing towards the grave told me that when he gets to Heaven—he is what they always call a *devout* Catholic and he has no anxiety whatsoever about his destination—and is asked by St Peter what, if he had lived longer, he would most wish to have prolonged, he will reply—"later afternoon golf".'

'That's right,' said my lawyer, 'you usually play around this time, don't you?'

I furtively leaned over to him while the ladies were still slurp-ing up the repulsive butter and lemon sauce. '*Around* this time,' I said with a meaningful ogle. '*By* this time I am usually recording my first par on the very difficult fourth hole whose green recedes invisibly into Long Island Sound.'

He is a guileless man—except in all matters pertaining to resid-uals and cassette rights—and he started as if a fox had started out of its burrow on our fast-eroding bank or bluff. 'Listen,' he whispered beseechingly, 'don't let us stop you . . . .'

I stopped him right there and begged the ladies to round things off with a liqueur. 'A dram of cognac, a soupçon of kirsch?' I suggested. 'Goodness, no thanks,'—thank goodness—they said.

'Really,' he implored.

'I tell you what,' I said, now in complete control of a situation not even imagined by the late, great Stephen Potter, 'we are going to dinner on the South Shore not far from you. Why doesn't Jane drive your ladies over an hour or so from now, *when they are ready to leave*? You and I will take your car, whisk around a fast eighteen holes at Island's End, and then we'll meet them all back at your place?'

I anticipated his next line, which was: 'You know, I'm ashamed to say I don't play, but I love to walk around. D'you think I could just amble along?'

The trajectory of our departure was never matched by Bugs Bunny. Fifteen seconds later we were on our way, leaving the women to surmise, 'Did they go for a swim? Irving doesn't really care for swimming.'

Island's End is a semi-public course, which means that at unpredictable times it is likely to be infested by a couple of busloads of the Associated Potato and Cauliflower Growers of Long Island. It was such a day. 'No way,' said the young pro, 'there are seventy-five of them out there and you wouldn't get through five holes before dark.'

'Too bad,' said my honest lawyer.

'I retain,' I said 'an escape clause for just such emergencies. I am a member at Noyac on the South Shore, and on Sundays not even guests can play.'

We snorted off to the ferry, trundled across the Bay, roared the five miles to the North Haven ferry, made it, and thundered up to the pro-shop at Noyac on the stroke of five. To all such jocularities as 'trying out a little bit of night golf, eh, Mr Cooke?' I turned a contemptuous ear. We were in an electric cart in one minute flat. (I hasten to say to snobs from the Surrey pine and sand country that no invention since the corn plaster or the electric toothbrush has brought greater balm to the extremities of the senior golfer than the golfmobile, a word that will have to do for want of a better.)

A natural and chronic modesty, which I have been trying to conquer for years, forbids my taking you stroke by flawless stroke through the following nine holes. ('I think we'll take the back nine,' I had muttered to my lawyer, 'it's a little more testing.') Suffice it to say that when we approached the dreaded eleventh—an interminable par five with a blind second shot up to a plateau three yards wide between two Grand Canyon bunkers—my friend was already goggling in the car or cart stammering out such memorable asides as, 'I don't believe it, it looks so goddam easy'.

An imperious drive had put me in the prescribed position for the perilous second shot. I was lining myself up, with the image of Nicklaus very vivid in my mind, when I noticed two large gorillas disturbing my peripheral vision. I paused and looked up in total, shivering control. Happily, on closer inspection, they turned out not to be gorillas but two typical young American golfers in their late twenties. They were about six feet three each, they carried the dark give-away tan that betokens the four-handicapper, and they waved at me nonchalantly and said, 'Go ahead, grandpa, we've got all the time in the world'.

Again I went through my casual Nicklaus motions. My three-wood followed the absolutely necessary arc and the ball came to rest precisely midway between the two bunkers. 'Many thanks,' I said briskly and waved back at the aghast gorillas. They marched fifty yards or so to what I was crestfallen to realise were their drives. Gallantly I indicated that no matter what the humiliation to me they should proceed. They banged two stout shots, one into the right bunker, the other on the edge of the pine woods.

By the time I came up to my ball, I was sitting upright with a commanding expression, worthy of Adolf Hitler arriving in Vienna in his hand-made Opel. The gorillas paused again, and waved again. 'Go ahead,' they said.

''Enks vemmuch,' I said. I took a five-iron with all the slow calculation of Geronimo Hogan. Ahead lay a great swale, another swirl of bunkers, a plateau and the bunkers guarding the sloping green. I heard in my ears the only sentence I have ever heard at the moment of address from George Heron, my old Scottish teacher of seventy-eight summers, springs and winters: 'Slow along the ground, big turn, hit out to me.'

It rose slowly like a gull sensing a reckless bluefish too close to

the surface, and then it dived relentlessly for the green, kicked and
stopped three feet short of the flag.

'Jesus!' cried one of the gorillas, 'd'you hit the ball this way all
the time?'

'Not,' I replied, 'since I left the tour.'

We left them gasping. And, to be truthful, I left the ninth hole
gasping.

'You never told me!' shouted my beloved friend and lawyer.

'I never knew,' I said.

ALASTAIR COOKE, *The Bedside Book of Golf,* 1965

The subtle (or not so subtle) interaction with partners and opponents is
another essential ingredient in the rich fabric of the game.

I have met strange opponents at golf; even stranger partners. I was
once paired with an elderly man, till then unknown to me, now
for ever memorable, who, after informing me, in confidence, that
he was a familiar figure at St Andrews in the gutta-percha days,
missed the object on the first tee, and, before I could intervene,
took a second swish at the ball, which hit the ladies' tee-box and
rebounded behind us into a gorse-bush. By weird methods we
reached the sixth tee, which is high up, with a valley on each side.
Off this he fell, to the left, while explaining the grandeurs of the
hole, and we became a three-ball match.

I have played in a two-ball sixsome, on the same side as an Irish
Protestant parson from Cork and an average-adjuster—a profession, surely, of magnificent vagueness to the uninitiated. I cannot
recollect who won. At Lindrick, that lovely course on the borders
of three counties, I have partaken in a one-hole, one-club match
with fourteen others, on a summer's evening, and the heavens were
darkened with balls, as was the sun, so they say, with arrows at
Marathon!

But the strangest man I ever knew on the links was a regular
opponent, of middle age and middle handicap. At once let it be
remembered that he was a man of kindliest character, which he
hid beneath fierce austerity of look and a power of invective that
were matchless in their time. His clubs seemed for ever to be on
the verge of total collapse; there was a driver from which string
and plaster flapped protestingly; a mongrel mid-iron, withered
and rusty, with which he played what he was pleased to call his

'push-shot'; a putter that was wry-necked and, though, I suppose, within the law, yet against equity and nature.

He would arrive, with erratic, curse-these-golf-club-garages swerves, on to the scene, give me a passing glare, disappear into the dressing-room, from which emerged sounds like a boot-fight between customer and assistant on a sale-day, then dash to the first tee. There he would allow me a curt nod of recognition, and shout such single comments as 'Off as soon as possible—late—course overcrowded—too many women—shouldn't play at all—talk to the secretary.' I have, indeed, known him to tee his ball in front of three waiting pairs, drive it off, and expect me to do the same. In others this would be accounted rude; in him it was natural, even magnificent.

As to his language during the hours of play, I have often thought that I missed the best; for he had known me since I was very young; sometimes, however, when he had sliced into a bunker (designed by the devil and executed by a brutish green committee), he would be heard to choke back some tremendous oath, composed of many moods and adjectives; and the word would roll back, stillborn, into the furnace of his frame! He regarded those playing around him as natural enemies and idiots, and once, when he missed a very short putt, he yelled down an adjoining fairway: 'D—n you, sir! Do you mind not flashing your brassy like that while I am putting?' On the tees during his practice swings the herbaceous vandalism was frightful to see; but, as the fids of turf spurted into the air, his comment would be: 'Club much too long—can't make it out,' or: 'Dropping the right shoulder again.'

Yet I have given, could only give, a faint and ghostly picture of this terrific golfer. Again, on a handicap of fourteen, a figure which facetious members would remark he had decided upon himself, as a Napoleonic committee of one, he was no mean opponent. I can see him now, at the end of the game, still in a fierce hurry, cranking up and maledicting his obsolete car, which he drove, on his more careless days, as wildly as his golf ball, perchance sweeping away in transit a tradesman's bicycle, and sending hens jittering and flapping through the hedges!

Of all types of opponent the most irritating is, I think, 'the convenient invalid,' that self-pitying monster whose hooks are the result of some galloping and mortal disease, who cannot putt because of hay-fever, who tops his mashies owing to synovitis in the elbow, is only playing because of medical advice, yet will, in fact, only reach his true form on the plains of Elysium, or elsewhere. I once knew such a fellow rudely cut short by a military opponent, who bawled at him: 'D—n you, sir; if I had your blasted lumbago, I could control my infernal swing, sir!'

Then there are those opponents who say 'Bad luck' when you miss from eighteen inches, others who remark: 'Funny, I thought you'd carried that bunker for certain,' or: 'Didn't think you'd miss that one, as my ball was giving you the line.' These, surely, must have their appointed after-life.

Less trying, but somewhat embarrassing, are those who say nothing at all, till suddenly, about the eleventh tee, they snap out 'One up' or 'Three down.' Others there are who, when you are four down at the turn, and there are five couples waiting to drive off on a bleak day, begin with that dispiriting sentence: 'Last time I played this hole . . .' and, I am afraid, we wish darkly that this could be the last time that he could play *any* hole. Ah! what bitter men are we golfers! How selfish in our thoughts! How petty in our little strifes! Day after day we wage our insignificant battles, curse our fate, as we smile with wan politeness; loathe our beastly opponent, his wretched, sniffing caddie, and his intolerable new set of irons. Grant me, when all is said, an old and forthright friend to play against, who swears and laughs at his own strokes and yours, likes to see you in a bunker and says so, seldom cries 'Good shot!' but, when he does, means it; takes golf as one of the world's eternal and necessary humours—and has a comfortable car!

R. C. ROBERTSON-GLASGOW, *Morning Post*

Michael Murphy is one of the founders of the Esalen Institute in California, and in the extraordinary book Golf in the Kingdom *he examines the connection between golf and spiritual discipline. Two chapter headings, which may put into context the following brief extract, are 'Relativity and the Fertile Void' and 'Universal Transparency and a Solid Place to Swing From'.*

The book is written as an account of a visit to Scotland and a meeting with Shivas Irons, the professional at the Burningbush Club in the Kingdom of Fife. After playing a round with Shivas and experiencing some of his unorthodox methods and theories, they have dinner with some of his friends. Afterwards each of the company explains what is for him or her the fascination of golf.

[17]

After the men have spoken the wife of the host gives her verdict:

'It's a way ye've found to get togither and yet maintain a proper distance. Yer not like women or Italians huggin' and embracin' each other. Ye need tae feel yer separate love. . . . All those gentle-manly rools, why, they're the proper rools of affection—all the wait-in' and oohin' and ahin' o'er yer shots, all the talk o' this one's drive and that one's putt and the other one's gorgeous swing—what is all that but love?

'Golf is for smellin' heather and cut grass and walkin' fast across the countryside and feelin' the wind and watchin' the sun go down and seein' your friends hit good shots and hittin' some yerself. It's love and it's feelin' the splendour o' this good world.'

MICHAEL MURPHY, *Golf in the Kingdom*, 1972

GOLF AS IT WAS

❖

Whatever the remote origins of golf it is clear that it was widely played in Scotland from at least the fifteenth century and that it remained for many years a notably democratic activity. However, the founding of the Honourable Company of Edinburgh Golfers in 1744 marked the beginning of one of the two great developments that have largely determined the shape of the modern game.

When the Honourable Company was followed ten years later by St Andrews, then by Blackheath in 1766 and Musselburgh in 1774, members' clubs were established for the first time with rules and the inevitable right to decide who should and who should not be allowed to join. With their special uniforms and club buttons they represented a significant move away from the merry free-for-all described by Smollett.

With Blackheath still very much a club for Scottish exiles in London it was over a hundred years before the first truly English club was set up at Westward Ho!, but after that the growth through the last quarter of the nineteenth century was enormously fast. No doubt much of this was due to the increasing prosperity and leisure of the middle classes at that time and also to the rapid development of the railway network which made travel easier for the population as a whole. But the other factor at work was the change in the ball.

The old 'featheries' cost half as much as a club, and long must have been the delays while they were looked for in the rough. They also quickly became unusable, especially in the wet. The invention of the 'gutty' and its introduction from 1848 onwards made golf substantially cheaper as well as easier to play, particularly for the beginner. It also marked the start of a fundamental change in the clubs which were carried. The leather-bound feathery would not stand up to the punishment of an iron club so that often, with the exception of a rut-iron, only wooden-headed clubs would be used. The gutty, on the other hand, actually flew better after its surface had been roughened, and the way was open for the development of the range of specialized irons that we know today.

*The arrival, in 1902, of the Haskell, the first rubber-cored ball, marked a
further significant step in making the game more accessible to the beginner
since, although it was again relatively expensive, it was much easier and
more comfortable to play with. Interestingly it does not seem as though either
the gutty or the rubber-cored ball in its early form actually flew further than
the feathery.*

Hard by, in the fields called the Links, the citizens of Edinburgh
divert themselves at a game called Golf, in which they use a curious kind of bats tipped with horn, and small elastic balls of leather,
stuffed with feathers, rather less than tennis balls, but of a much
harder consistence. These they strike with such force and dexter-
ity from one hole to another, that they will fly to an incredible
distance. Of this diversion the Scots are so fond, that, when the
weather will permit, you may see a multitude of all ranks, from
the senator of justice to the lowest tradesman, mingled together,
in their shirts, and following the balls with the utmost eagerness.
Among others, I was shown one particular set of golfers, the
youngest of whom was turned of four-score. They were all gentlemen of independent fortunes, who had amused themselves with
this pastime for the best part of a century, without having ever felt
the least alarm from sickness or disgust; and they never went to
bed without having each the best part of a gallon of claret in his
belly. Such uninterrupted exercise, co-operating with the keen air
from the sea, must, without all doubt, keep the appetite always on
edge, and steel the constitution against all the common attacks of
distemper. TOBIAS SMOLLETT, *Humphry Clinker*, 1771

*Apparently known as 'a mighty swiper', i.e. driver of the ball, Dr Carlyle
clearly had a good opinion of his own ability. It is interesting that while it
was only the Scots who knew how to play, there was already an established
golfing ground at Molesley Hurst. No trace of it now remains.*

Garrick was so friendly to John Home that he gave a dinner to
his friends and companions at his house at Hampton, which he
did but seldom. He had told us to bring golf clubs and balls that
we might play at that game at Molesey Hurst. We accordingly set
out in good time, six of us in a landau. As we passed through
Kensington the Coldstream Regiment were changing guard, and,
on seeing our clubs, they gave us three cheers in honour of a diversion peculiar to Scotland; so much does the remembrance of one's

native country dilate the heart when one has been some time absent. . . . Immediately after we arrived we crossed the river to the golfing ground, which was very good. None of the company could play but John Home and myself and Parson Black from Aberdeen.

Garrick had built a handsome temple, with a statue of Shakespeare in it, in his lower garden on the banks of the Thames, which was separated from the upper one by a high road, under which there was an archway which united the two gardens. . . . Having observed a green mound in the garden opposite the archway, I said to our landlord, that while the servants were preparing the collation in the temple I would surprise him with a stroke at the golf, as I should drive the ball through his archway into the Thames once in three strokes. I had measured the distance with my eye in walking about the garden, and, accordingly, at the second stroke, made the ball alight in the mouth of the gateway and roll down the green slope into the river. This was so dexterous that he was quite surprised, and begged the club of me by which such a feat had been performed. ALEXANDER CARLYLE, *Autobiography*, 1860

THE INTRODUCTION OF THE GUTTY
Golf was rendered expensive in those days, not by the clubs, which were cheaper then than now, but by the balls. Their prime cost was high, and their durability not great. On a wet day, for example, a ball soon became soaked, soft and flabby; so that a new one had to be used at every hole in a match of any importance. Or, on the other hand, a 'top' by an iron in a bunker might cut it through. This I have frequently seen occur.

The making of first-class feather balls was almost a science. For the benefit of the uninitiated, I shall endeavour to explain the operation. The leather was of untanned bull's hide. Two round pieces for the ends, and a stripe for the middle were cut to suit the weight wanted. These were properly shaped, after being sufficiently softened, and firmly sewed together—a small hole being of course left, through which the feathers might be afterwards inserted. But, before stuffing, it was through this little hole that the leather itself had to be turned outside in, so that the seams should be inside—an operation not without difficulty. The skin was then placed in a cup-shaped stand (the worker having the feathers in an apron in front of him), and the actual stuffing done with a crutch-handled steel rod, which the maker placed under his arm. And very hard work, I may add, it was. Thereafter, the aperture was closed, and firmly sewed up; and this outside seam was the only one visible. When I say this, I of course refer to balls when new. Veterans showed the effects of service in open seams, with feathers out-looking; and on a wet day the water could be seen driven off in showers from a circle of protruding feathers, as from a spray-producer. A ball perhaps started a 'twenty-eight', and ended a forty pounder.

The introduction of gutta-percha balls effected a complete re-volution. Their cost was small, their durability great. I believe I may with justice claim the credit of having first brought them to the notice of the golfing world, and this at the Spring Meeting of the Innerleven Club in 1848. The previous month, when on my way home from a two years' stay in France (where, by the way, golf was then unknown), I chanced to see in the window of a shop down a stair in St David Street, Edinburgh, a placard bearing the words—'New golf balls for sale.'

I found them different from anything I had seen before; and was told by the shopman they were 'guttie-perkies'.

'Guttie-perkie! What's that?' I asked; for I had never heard of it.

'It's a kin'o' gum like indiarubber.'

'What kind of balls does it make?'

'I ken naething aboot that—best try yin yoursel'!'

I bought one for a shilling. It was not painted, but covered with a sort of 'size', which, after some practice with my brother James, who was a good golfer, I saw reason to scrape off.

I then determined to try it upon the Innerleven Links, against Mr David Wallace, a golfer with whom I often played, and who always beat me. I noticed that after I had 'teed', he looked at my ball with great curiosity; so I told him its history and the result of my experiments; and away we went. The upshot of our day's play was, that I beat him by thirteen holes—a thrashing, he said, such as he had never had in his life. However, he, too, soon took to the 'guttas'; and many a beating he gave me afterwards. Still, I was much more on an equality with him than before.

I won the silver medal against him in April, 1848; and it was at that meeting I showed the new ball to Allan Robertson and Tom Morris. It was the first time either had ever seen a 'gutta'. I told them of its great superiority to 'feathers', and that the days of the latter were numbered; but Allan would not believe it. At my request, he tried the new ball; but instead of hitting it fairly, struck it hard on the top in a way to make it duck (which, by the way, no one could do more deftly than Allan).

'Bah!' he said; 'that thing'll never flee!'

I, however, struck it fairly, and, to Allan's disgust, away it flew beautifully!

For a long time Allan persisted in this opposition to 'guttas'. He has often told me, when I wanted him as partner in a foursome, that he would not play unless I used feather balls—a condition to which I, of course, acceded. At last, however, even Allan had to yield. He not only began to make them (as many others had by that time done): but played with them.

Tom Morris, on the other hand, took the whole thing in a different way. His customers informed him that 'feathers' were doomed; he at once made 'guttas', and very successfully. Nay, if I remember rightly, his difference with Allan on this subject led to their separation.

For long I made my own balls, and at small cost. The only point in which 'guttas' were at a disadvantage, as compared with 'feathers', was that they did not hold their course well in high wind, specially a side one. After some scheming and experiment, my brother and I succeeded in inserting and fixing lead securely in the centre of the ball, so that it putted accurately.

Nearly all the medals I gained were won with leaded balls; and I used them regularly until my stock was exhausted. (The making of them ceased at my brother's death.) They were well known at that time; and when I played at St Andrews with Hugh Philp (a good player and deadly at the short game) he used to ask me for one of my leaded balls.

They were, however, severe upon clubs—the fairest struck ball often breaking the head through the centre. Many of Philp's fine clubs have I broken in this way; and when I complained of rotten wood, he would answer: 'Hoo the deevil can a man make clubs to stand against lead?'

Other players, again, used to lead their balls by rolling them when warm in lead filings; but as these were on the surface, they fell out when the ball was struck, and gave it a very unsightly appearance. That plan was inferior to ours.

[23]

I may mention that for a considerable time I played with unpainted balls under the impression that they flew better; but there was, of course, the draw-back that they were difficult to find.

H. THOMAS PETER, *Golfing Reminiscences by an Old Hand*, 1890

WESTWARD HO!

I can remember, on the question of the modern cost of golf at Westward Ho! that almost the sole expense was for clubs and balls. Nature, and a few sheep cropping the grass, which never grew long on that link's turf, did all the rest for us. I am not quite sure that we did not think we had done a very big thing—almost gone too far on the lines of luxury and precise attention to having all in perfect order—when one of the members sent to Scotland for a hole cutter.

Previous to that enormous piece of extravagance the mode had been to cut out the holes with an ancient dinner-knife, or clasp knife. Sometimes—but even this seemed a first lapse towards a precision that was almost meticulous—a gallipot was held on the ground to give a circle round which the knife should incise the turf. The holes grew and grew, in diameter, with wear, and were notched away here and there at their edges, although there was no large multitude of golfers to do the wearing away, so that a wily man, finding the putt a difficult one with a troublous little rise in the ground interposing in the direct route between him and the hole, would putt so as to try to let the ball fall in by one of those bays or notches in the edge if there were an easier putt towards the kindly welcoming entry.

There was no fixed day for the cutting of a new hole, but when the leading players on any day found one of the holes too far gone in decay to be longer endurable, he would out with his knife and cut a new hole to which he and his opponent would putt, and thereafter all who followed could putt likewise.

The site of the hole would be indicated by the feather of a rook or a gull stuck into the ground at its edge, for by these simple means did these good men, our forbears, mark the situation of the holes.

In some old rules of golf—remarkably few in number compared with the elaborate code of today—you may find a provision that the ball shall be teed for the next hole not less than three, nor more than four, club lengths from the hole lately played out.

That gives you an indication of the state of original sin in which the putting greens were left, when you could do no harm to them so long as you did not tee up and take your stance and drive off nearer the hole than three club lengths.

It indicates, too, and rightly, that there were no set teeing marks. Distance from the previous hole, as stated, gave you the approximate limits within which you might tee. And as there were no teeing marks, neither were there tee boxes for sand.

The easiest and natural and usual way to get sand for the teeing of the ball was to scoop it out of the last hole played and this was the plan always followed. It had the effect, if the turf was sound so that the hole kept its edges good and was not shifted for some while, that as more and more sand was taken out the hole grew more and more profound, until it often happened that one had to lie down so as to stretch one's arm at full length in order to reach the ball at the bottom of the hole.

Once upon a time there was a rabbit which played a very remarkable practical joke upon some of us at Westward Ho! With great discrimination it refrained from any onset upon the turf in the neighbourhood of the hole which might have given warning of its presence, but made use of the hole itself as an entry providentially constructed, and dug down from the bottom of the hole nobody ever discovered how far. Of course, the result was that as party after party came to the hole they putted out but were altogether unable to retrieve the ball which had gone down into the very bowels of the earth.

It would, to be sure, have been only right and humane of the first party to whom this disaster happened to block up the floor of the hole so that the same should not befall those who came after. But each party in turn, having lost their own balls, did not see why they should go back to be laughed at by the others. Accordingly, they left the trap gaping open so that all should be taken in equally and none have the laugh over the other.

But the rabbit must have laughed at the bottom of its burrow, and probably set up a gutta-percha store on the result.

HORACE HUTCHINSON, 'Old Memories of Westward Ho!',
The Midland Golfer, 1914

The association of golf with bad language is a frequently recurring theme.

At Westward Ho! . . . we had many old Indian Officers, with livers
a little touched, and manners acquired in a course of years of deal-
ing with the mild Hindoo, and because the golf ball would not
obey their wishes with the same docility as the obedient Oriental,
they addressed it with many strange British words which I delighted to hear and yet stranger words in Hindustani which I much
regretted not to understand. But a sight that has been seen at
Westward Ho! is that of a gallant Colonel stripping himself to the
state in which Nature gave him to an admiring world, picking his
way daintily with unshod feet over the great boulders of the Pebble
Ridge, and when he came to the sea, wading out as far as possible, and hurling forth, one after the other, beyond the line of the
farthest breakers the whole set of his offending clubs. That the
waves and the tide were sure to bring them in again, to the delight
of the salvaging caddies, made no matter to him. From him they
were gone for ever and his soul was at rest.

Of course he bought a new set on the morrow, so it was all
good for trade and Johnny Allan. It also afforded a splendid spectacle to an admiring gallery. Really we have lost much at Westward
Ho!, even if we have gained much, by the bringing of the Club-
house across the common. It was delightful, after golf or between
the rounds, to bathe off that Ridge, or sit on it and watch the sea
tumbling. . . .

HORACE HUTCHINSON, 'Fifty Years of Golf', Country Life Library, 1914

On a day in April, I walked round the Links with a 'foursome': the
only time I ever did so. It is sad to make such a confession: but
truth must be told. My brother Alexander and Lord Colin Campbell
played against Tulloch and a golfer departed. It was extraordinary
how peppery the golfers became. Tulloch and his partner were
being badly beaten, and became demoralised. Tulloch, seeing his
partner doing something stupid, made some suggestion to him. On
which his irate friend brandished his club in the air, and literally
yelled out, 'No directions! I'll take no directions!' Tulloch used to
complain that an old story of the Links and their provocations,
applicable to another Principal, had come to be told of him. 'How
is the Principal getting on with his game?' was asked of one of the

caddies of a returning party. 'Ah!' said the caddie, with an awestrick-
en face, 'he's tappin' his ba's, and dammin' awfu'.'

A. H. K. BOYD, *Twenty-Five Years of St Andrews*, 1892

*Anthony Trollope upset his hosts by his coarse manners at dinner and seems
to have been equally unruly on the links.*

Later in the season, Anthony Trollope and his wife paid their first
visit to this place. They staid at Strathtyrum with John Blackwood.
Trollope tried to play golf. It is a silent game, by long tradition:
but Trollope's voice was heard all over the Links. One day, having
made a somewhat worse stroke than usual, he fainted with grief,
and fell down upon the green. He had not adverted to the fact that
he had a golf-ball in his pocket: and falling upon that ball, he started up with a yell of agony, quite unfeigned.   Ibid.

*Captain Molesworth was one of the great characters—and evidently one of
the great competitors—of the early days at Westward Ho! He used, as this
extract makes clear, only three clubs: driver, iron, and putter, which were
known as Faith, Hope, and Charity respectively.*

On Wednesday, Sept. 5, Capt. Molesworth played a long-talked-of
match made at the Whitsun meeting, in which considerable interest was taken. The terms were as follows: Capt. Molesworth to
walk to the links (a distance of three miles) and play on any day
six rounds of golf, carry his own clubs, and walk back between
daylight and dark, and do the six rounds under 660 strokes. The
Captain left home at 5.20 a.m., reached the Iron Hut at six o'clock,
and commenced to play at 6.10. The dew was so thick on the grass
that it was as bad as playing in the heaviest rain, as far as wood-
en clubs were concerned, and the face of the driver was soon
driven in, and at the twelfth hole went all to pieces. The captain
came to grief at the ninth hole, taking 14 strokes, and played six
holes in with his iron and putter, taking 120 strokes for the first
round; the second was played with nothing but irons, and accomplished in 105 strokes; in the third the captain again came to grief
at the ninth hole, taking 15 strokes, ending the round in 122 strokes;
the fourth in 108; the fifth, 102; sixth 105. As these six rounds came
to 662 strokes, Capt. Molesworth played a seventh round, which
he did in 104 strokes, thus doing six rounds in one day, without
counting the first round, in 646 strokes, winning by 14 strokes upon

the six rounds; walked back, and reached home at twenty minutes to seven p.m. To play the round takes from one hour and thirty-five minutes to one hour forty-five minutes' hard walking. Capt. Molesworth played the first three rounds without stopping and, with two intervals of five minutes each, accomplished the seven rounds by half past five. The question arises whether Captain Molesworth, having played six rounds in two strokes over the number, viz., 662, was entitled to play another round. Capt. Molesworth's backers say that the match was to play six rounds in one day, between the hours of daylight and dark: and, therefore, it did not matter how many rounds he played, if six rounds were played in under 660 strokes, and that he was entitled to leave out the first or any other round, so long as six whole rounds were done under the number. The other side contended that, as six rounds were played, and the number of strokes taken was over 660, the match was lost.

A case will be drawn, and the matter referred.

The Field, September 1877

❖

A BAD DAY AT WESTWARD HO!

Last week from time to time there was a good deal of bunker running. The squalls, though short, were heavy; at the first symptoms of dropping there was a cry of 'here it comes' and the whole four-somes with their caddies, skeddaddled for the nearest bunker, and stretched at full length under the ledge till the fury had passed away. In the distance as the storm drove to leeward, fellows who were not near bunkers made for big rushes, which with the help of an umbrella, don't form bad shelter; others regardless of squalls, took the opportunity of passing their leaders. On the 11th the rain which had before come at intervals, gave up and came down regular; the squalls ceased as the gale took up the blowing all to itself.

Newspaper report, 19 August 1874

Obviously the weather took a turn for the better after the last report, and the players were able to demonstrate their skills.

RISING STANDARDS

Never since the white globe commenced to traverse the Westward Ho! links has golf been so continuous as during the last few weeks and much fine golf has been discoursed. Some have supposed that the links are easier than heretofore but not so. The averages 96, 98 to 100 are not that the links have given up their bunkers or rushes, but that the science of golf is becoming more familiar to students of the noble game; perhaps more than in any other links is the cultivation of heavy iron approach strokes needed, and so accurate have these become in the measurement of these distances that the deadly fall of the lofted ball strikes terror into the peaceful mind of the stranger who believes after the second drive he stands in as good a position as his adversary.

Newspaper report, 5 September 1874

❖

'ALL DOWN'.

To the Editor of GOLF.

SIR,—As an orthodox golfer, born and bred in the true faith north of the Tweed, though now unhappily chained to London, I have often been surprised and pained, not so much at the Englishman's madness for the game, as at the manner of it.

To him we owe, amongst other things, the perplexing and uncalled-for mutilation of the ancient rules to suit the exigencies of local greens, the invention of fixed match-play odds, 'Colonel Bogey,' and the electro plated cruet-stand, with Golf-ball holders, and supported by miniature Golf clubs, suitable as a prize for handicap competitions. And now, when we are battling with these things, with the horrible spectre of County Golf, like the Disestablishment of the Church, looming in the distance, when our irons are rattling in their sockets by reason of the hard ground, and the putting-greens are like a parade-ground after an earthquake, a worse thing has come upon us.

What do you think happened to me the other day?

I was playing a match with a man, and, at a point when the game became square, he said to me, in the most matter-of-fact way in the world, 'All down!'

'What?' said I, thinking I had not heard aright. 'All square; is it not?'

'Yes,' says he; 'all down!'

Since then I have encountered the expression pretty frequently on various greens in the neighbourhood of London, and I believe it is coming into general use.

Golf, as she is played nowadays, is too often a sad draggle-tail;

but, if 'all down,' for 'all even' or 'all square,' is a specimen of Golf as she is going to be spoke, it will, I fear soon be 'all up' with golfers who have any sensibility left.

I am, Sir, &c.,

G. G. S.

Golf

❖

SOME EARLY GOLF BALLS

On top of my interest in the game itself, I took a tremendous interest in the clubs and balls, particularly the latter. I had seen many changes in the golf ball, and I believe the great development in the game of golf is directly attributable to the wonderful strides made by the manufacturers in perfecting the ball and making the game a more pleasant one to play. As I have already said, the old hard and solid gutta-percha ball was succeeded by the rubber-filled Haskell. Then came the Kempshall Flyer. These balls were extremely resilient, but their qualities of durability were far from satisfactory.

I have seen both of these crack wide open and become unfit for play on the first or second shot. If the golfer topped one of them, it was the finish. They just would not last at all. Then came a ball called the Spalding Wizard. But, like the first two, it could not stand up under punishment. British-made balls put in an appearance and they were easily destroyed.

About 1906, the Goodyear people, who had helped Doctor Haskell in perfecting the original rubber-cored ball, placed on the market a ball called the Pneumatic. This ball had a rubber cover or shell into which air was compressed. It caught the fancy of the golfers for a time, but the Pneumatic had to be hit powerfully in order to get it along, and players in 1906 were new at the game and did not have the necessary hitting powers. The younger element had fair success with it, because they could hit harder and therefore got the better results. It lacked the resiliency of the rubber-cored balls, and although it was fairly durable in so far as cutting the cover was concerned, it had other faults which were distressing. Many times it was knocked out of shape, and when this did not occur it exploded either in flight or in the caddie bags. Golfers had many interesting experiences with the Pneumatic.

With a view to improving the ball and making it indestructible, the Goodyear firm introduced in 1907 a ball called the Silk Pneumatic. It was constructed in this fashion: the cover was made of rubber, and inside were woven silken strands something along the idea of the motor-tyre fabric. This strengthened the cover materially. Inside this was placed a substance resembling gelatine, in solid form and about a quarter of an inch thick. The gelatine layer fitted into the inner cover or shell. Then the air was pressed inside, and you had your ball. It was a big improvement over the Pneumatic, seemed a trifle more lively, and stood up reasonably well. Alex Campbell played the ball, as did Alex Smith and other well-known professionals of that time.

In 1907, Campbell was playing in the Open Championship at the Philadelphia Cricket Club and had a fine chance to win. He played two shots with his Silk Pneumatic to the edge of a green and took four more to hole out from about twenty feet. Picking his ball out of the cup, he discovered it had burst, the air had escaped, and he had been playing those four shots with nothing more or less than a mushy bag. Those Pneumatic balls played better after they had been hammered round quite a bit, but there was always the danger that something would happen, and they were thrown into disuse.

In 1908 liquid-filled balls were made. Spalding put out a ball called the Aqua. Another was named Water Core. Before these innovations, the centre of a golf ball consisted of a small piece of gutta-percha the size of a marble, and around this were wound the rubber or elastic bands. Balls with liquid centres were very popular, but manufacturers had much trouble making them durable. Nevertheless, they were far superior to the gutta.

Nowadays golf balls are made in two styles—either recessed or mesh-marked. With an exception here or there, all rubber-cored balls up to 1908 or 1909 had the bramble or pimple marking. I think it was in 1908 that this marking was invented and became known as the Dimple. The Glory Dimple, with its red, white, and blue decoration, was a grand ball. And one more thing about golf balls: there has been so much comment in recent years about the size and weight of the golf ball. Those early rubber-cored balls were very light, floated in water, and no one gave a single thought to size or weight!   FRANCIS OUIMET, *A Game of Golf*, 1933

❖

LAMBERHURST

The worst thing about Lamberhurst golf was that it provided very poor practice for playing anywhere else. In fact one could almost say that it was 'a game of its own'. For one thing, you were perpetually hoicking the ball out of tussocky lies; and for another, the greens had justifiably been compared to the proverbial postage-stamp. If you pitched adroitly on to a green, it was more than likely that you wouldn't remain there. If, on the other hand, your ball fell short, you stopped where you were, which was in the rough grass. And the otherwise almost hazardless charm of our local links didn't always atone for these disadvantages, especially when one happened to be playing a medal-round at the Spring or Autumn Meeting. During the summer months the course got completely out of control and nobody bothered to play there except Squire Morland himself, and he had seldom done the nine holes in under fifty at the best of times. Go there on a fine April day, however, and there was nothing to complain of, provided that one gave the idyllic pastoral surroundings their due and didn't worry about the quality of the golf. I say 'pastoral' because the place was much frequented by sheep, and I cannot visualize it without an accompaniment of bells and baa-ings.

Standing near the quiet-flowing tree-shaded river at the foot of the park, one watches a pottering little group of golfers moving deliberately down the south-westerly slope. It is one of those after-luncheon foursomes in which the Squire delighted; and there he is, playing an approach-shot to the third hole in that cautious, angular, and automatic style of his. The surly black retriever is at his heels, and his golf-bag has a prop to it, so as to save him stooping to pick it up, and also to keep his clubs dry. The clock on the village school strikes three, and one is aware of the odour of beer-making from the Brewery. The long hole to the farthest corner of the park is known as 'the Brewery Hole'. And now they are all on the green, and gallant old General Fitzhugh, who had conspicuously distinguished himself in the Afghanistan campaign some thirty years before, is taking tremendous pains over his put. The General has quite lately acquired one of those new Schenectady putters, mallet-shaped and made of aluminium, and popularized by Walter Travis, the first American who ever reached and won the Final of the Amateur Championship in England; and the non-success of his stroke is duly notified when he brandishes the weapon distractedly above his head. I now identify the stocky upright figure of my old friend Captain Ruxton, who evidently has 'that for it', and sinks the ball with airy unconcern; whereupon the Squire, I can safely assume, ejaculates 'My word, that's a hot 'un, Farmer!' in his customary clipped and idiomatic manner.

The fourth member of the party, I observe—unless one includes a diminutive boy from the village who staggers under the General's bristling armoury of clubs—is Mr Watson, a tall, spectacled Scotsman, still in the prime of life, whose game is a good deal above Lamberhurst standards. Watson is a man well liked by everyone—without his ever saying much, possibly because he can't think of anything to say. My mother once remarked that when Mr Watson ran out of small talk at a tea-party he told her that he always gave his hens salad-oil for the good of their health. But his favourite conversational opening was 'Have you been to Macrihanish?'—Macrihanish being an admirable but rather un-get-at-able golf course within easy reach of the Mull of Kintyre. A person of strict principles, he had never been heard to utter the mildest of expletives, even when he found one of his finest drives reposing in the footprint of a sheep. After making a bad shot he used to relieve his feelings, while marching briskly toward his ball, with a snatch of cheerful song. 'Trol de rol de rol' went Mr Watson. But one had to have been to Macrihanish if one wanted to get much out of him in the way of conversation.

The friendly foursome is now well on toward the fourth green, where—in my mind's eye—I am standing by the tin flag, which requires repainting and has been there ever since the Club was founded. Following the flight of their tee-shots, I have remembered with amusement that Squire Morland occupies what might be called a duplicated position in the realms of print. His name appears, of course, in Burke's *Landed Gentry* and similar publications of lesser importance. But it also figures in *The Golfing Annual* as 'green-keeper to the Lamberhurst Golf Club'. (Nine holes. Subscription 21/- per annum.) The Squire's assumption of this sinecure appointment is due to the fact that by so doing he cannily obtains for himself and his cronies all the golf balls that he needs, at wholesale price. It is just conceivable that he does pull the light roller up and down about once a year; but I never heard of him doing so, though I myself had put in an hour's voluntary worm-cast sweeping now and again—the green-man being rather apt to neglect his duties.

[33]

In the meantime the foursome has another fourteen holes to play before it adjourns to the House for tea and a stroll in the garden to admire the daffodils. And while those kindly ghosts gather round me on the green I can do no more than wish that I could be greeting them there again, on some warm April afternoon, with the sheep munching unconcernedly and the course—as the Squire used to say—'in awful good condition'. But as the scene withdraws and grows dim I hear a blackbird warbling from the orchard on the other side of the river; and I know that his song on the springtime air is making even elderly country gentlemen say to themselves that one is only as old as one feels, especially when there is nothing forbidding about the foreground of the future—into which they are now, with leisurely solemnity, making ready to smite the ball.
SIEGFRIED SASSOON, *The Weald of Youth*, 1942

Lamberhurst is now a thriving club with eighteen holes of immaculate fairways and certainly no sheep. The brewery has gone, too, but the Morland family still live in the house and a tablet on the wall of the church behind the fourth tee commemorates the old squire.

LOMBARTZYDE, CHRISTMAS 1914

My brother Edward, who had retired from the Army and was farming in Virginia when war broke out, was now in a New Army battalion of his old regiment, the South Staffords. He came out and spent Christmas with me and had his first glimpse of the war. There was a well-known golf course at Lombartzyde on which the Belgian championship had been played the year before, so after a good lunch, washed down by Nuits St Georges, Ted and I who used to have very well-contested matches, set out on Christmas Day to try and play a round. We found the club-maker's house, now a strong-point in the front line, still partly standing, and plenty of clubs and balls lying about. Thus equipped we managed to play a good match of seven or eight holes, though the number of bunkers had become rather excessive. We had to hurry on one or two greens owing to the unsporting conduct of snipers and spent half an hour in a pot bunker guarding the sixth hole. This was in contrast to the procedure of the Bulgarians on the Struma, who allowed our troops to play football under their guns without molestation. The match ended all square and one to play, which we decided to renounce as the Germans seemed to be getting cross.
TOM BRIDGES, *Alarms and Excursions*, 1938

There seems, at least in retrospect, a wonderful, carefree innocence about the early days of the tour, before the pros discovered that they needed managers and PR people and a permanent staff to travel with them.

L.A. . . . lured the largest galleries of the year, crowds sparkling with movie celebrities and rising starlets. There were so many one year, in fact, that Dick Metz had to park two miles from the course before the second round and buy a ticket to get in. This would not have been so embarrassing for the sponsors if Metz hadn't been leading the tournament at the time.

'We knew everybody in Hollywood,' says Jimmy Demaret. 'It was pretty impressive to be hanging around Bing Crosby and folks like that all the time. And there were an awful lot of dandy little old gals around. We didn't know who they were. They had different names then. But we realized later that they were Susan Hayward and that kind of thing.'

The tour might stay on the Coast for a month. It would go from L.A. to a Riverside Open, where the players would stuff their golf bags with fruit from the citrus trees in the rough, or it might go to a Pasadena Open, where they scooped niblicks around the Rose Bowl. It might go to an Agua Caliente Open. And one they certainly didn't miss was the Bing Crosby Pro-Am at Rancho Santa Fe, which was only thirty-six holes. It would also head north to a weird thing called the San Francisco Match Play championship. The pros would get into the Sir Francis Drake Hotel, dine at Vanessi's and Nujoe's with a wealthy host and work up the pairings in the locker room. 'You figured out who you thought you could beat and challenged him,' Demaret says. 'And you hoped you didn't get Leonard Dodson because he'd pay a guy to follow you around with a camera and click it on your backswing.'

After the flamboyant times on the Coast, fan belts permitting, the tour moved lazily through the Southwest, the South, the East and the Midwest, until, quite sensibly, it came to a dawdling end as football season began. . . .

It was not always easy getting from one place to another. Hogan and Nelson would be more intent on reaching a destination than some of the others because they wanted to practice. They would form a mini-motorcade, Byron and Louise following Ben and Valerie, and dart for shortcuts like Bonnie and Clyde. 'Ben wanted to get there and start hittin' golf balls,' Demaret says, with a slight trace of bewilderment. 'Nobody ever practiced. First guy I ever saw do it was Bobby Cruickshank, about 1932. I didn't know what he was up to. But Ben *really* practiced. I'm sure he invented this business of spending two and three hours hitting balls.'

Demaret, of course, would be late because he would still be throwing a party from the previous week. And Lloyd Mangrum, with Buck White or Leonard Dodson, would have found a card game—seven-card low, pitch, auction bridge, casino—while Sam Snead would have to sit around and wait for them in the car. Johnny Bulla's car. 'For a while, we traveled in Bulla's Ford sedan. Can you imagine anyone loaning Lloyd Mangrum, Sam Snead and Buck White his car?' Mangrum says. Some cars made the next stop easily, like Walter Hagen's front-wheel drive Cord, or Denny Shute's Graham-Paige. But others didn't. Johnny Revolta has vivid memories of a fan belt going out somewhere in the infinity of a Texas horizon. His companion at the time, Henry Picard, sat in the car and held his head while Revolta hiked twelve miles, got rescued by a Mexican family, finally found a belt to fit and, six hours later, returned.

This would have been on the way to San Antonio for the Texas Open, the oldest of the winter events (1922). The tournament was distinctive for a number of reasons, not the least of which was that it was usually played among the willows and pecans of Brackenridge Park, a short, tight little public course featuring rubber mats for teeing areas. It was on this course that the most headlined practice round of the era was played one year. In a group with Hogan, Nelson and Paul Runyan, the easy-going Jug McSpaden, wearing sun goggles as always and putting with a mallethead, shot a stunning 59. The round was accented by the fact that Nelson had a 63, Hogan a 66 and Runyan a 71. And it was also at Brackenridge Park one year that Wild Bill Melhorn, who made most of his expenses at the bridge tables, finished with a seventy-two-hole score that looked as if it might win, provided Bobby Cruickshank didn't par the last hole. Melhorn climbed a live oak and tried to heckle Cruickshank into a three-putt on the 18th, but when Bobby survived, Melhorn toppled off the branch and into the crowd, without backspin.

From San Antonio, the tour normally frolicked onward to New Orleans, which was both a good and bad town for a riverboat gambler like Lloyd Mangrum. He once arrived on Mardi Gras Eve with the town so crowded—and himself so busted—that he joyfully bunked in the city jail for two days. 'The only hard part was going without cigarettes,' says Mangrum. 'Try that course sometime.'

Mangrum went on the tour with a $250 stake he got from John Boles, the singer. Cards kept him on it when his golf didn't. 'I never won as much as everybody thought at the card tables,' he says. 'But the first time I ever got enough money to afford it, I bought me a big LaSalle with a couple of four-foot chrome horns on the front fenders, and folks thought I was rich.' Stories would spread on him. And grow. When he won $4000 at gin rummy in Arizona it would be $25,000 by the time it reached Toots Shor's in New York. 'The public thought you got thousands from the club companies,' he says. 'But all I know is I thought it was a big lick to get $300 and some clubs and balls out of MacGregor.'

There was usually a way to get something out of somebody. If Mangrum couldn't do it with cards, Leonard Dodson would think of something else. 'We're in Niagara Falls, I think it was,' says Lloyd, 'and I've got about $1.35 and Dodson's got $1.25. Together we can't buy enough gas to get half-way where we're going. So Dodson scares up this Canadian millionaire and bets him $500 he can outrun his automobile at 100 yards. Now we can't get $500 up if we rob everybody on the tour, and I don't think Dodson can outrun me if I got a heart attack. But Leonard takes the guy out in the country to an old dusty farm road where the tires can't get any traction in the deep, soft sand, and damned if Leonard doesn't win the race on his start.'

Potential sponsors for all of these tournaments were mostly where Fred Corcoran could find them, and they came in all shapes and sizes and bankrolls. A sponsor was usually the first to leap in front of a camera, putting his arm around a Snead, a Hogan or a Nelson, buddy-buddy. But not all of them were like that in Corcoran's day. There was this one gentleman, quite amiable, who put up a good portion of cash for the 108-hole Westchester Classic in 1938. Except no photographs, please. Corcoran himself had to present Snead with the victory check so the sponsor could hide from the press. Four years later everyone discovered why. The gentleman was convicted of printing bogus whiskey labels. And no one has yet, to anyone's knowledge, tracked down one of the sponsors of the Indianapolis Open of 1935. It took Al Espinosa, the winner, three years to get his thousand dollars prize money from the PGA because the Indianapolis fellow skipped town with the purse.

In all there were no more than twenty-five tournaments a year, compared to forty-five today. If the total prize money added up to $150,000, that was sensational. It looked as good as the $5.5 million of 1968. And a player in the Thirties knew that if he could reach most of those tournaments, if he could attach the clubheads, which were made in Scotland, to the shafts, that were made in Tennessee, if he could assemble a set he liked, and if he could survive the card games and the parties and the fan belts, he could bank as much as $6000 for the year and rate as high as fifth on the money list.   DAN JENKINS, *The Dogged Victims of Inexorable Fate*, 1970

MASTERING THE ART

❖

No player needs to be reminded that golf is a difficult game. We demonstrate that fact to ourselves and our partners every weekend. Even the
professionals, with their apparently immaculate swings, are constantly
reported to be seeking help from their own particular gurus, and even they
can go from a 64 to a 74 on two successive days on the same course.

That this difficulty is an essential part of the fascination of the game is
clear. The old joke that hell is where you hit a perfect shot every time has
its truth— though all of us would settle for a weekend pass from time to time.

The game is not as easy as it seems. In the first place, the terrible
inertia of the ball must be overcome.

LORD WELLWOOD, *Golf*, 1895

⚓

DISAPPOINTMENT

My feeling is that while 'Golf' is the name by which we know the
game what we are actually playing should be called 'Disappointment'.

Consider the facts. First, let us look at our physical attributes.
Most golfers have the regulation four limbs, standard torsos with
heads attached and a pair of eyes, equally spaced. While this combination makes for symmetry and allows us to perform ordinary
tasks, such as climbing Everest, when it comes to hitting a golf ball
it is useless.

Second, let us consider the equipment we have to use. A more
ill-designed set of implements could hardly have been created. Yet
we descend upon each new model with cries of glee, drooling over
the gleaming heads.

What deluded fools we are. When one looks at the size of the
golf ball and then looks at the area allocated to hit that small
sphere, then it is remarkable that contact between the two is ever
properly achieved.

Such is our foolishness, however, that we not only expect to make contact but we also expect to make a contact that will send the ball (a) up into the air (b) a long way and (c) in the direction we have chosen. All this we expect to achieve without even considering (d) which dictates that, in order to reach these expectations, we have to make the club travel in a full 360 degree arc.

Thus, when we step on to the tee in full possession of our symmetrical arms, legs, muscles and joints and combine these with the aforementioned hardware we are primed for disappointment. It is at this blending stage that we get the first inkling that all is not as it should be for, in order to move the club in the prescribed arc, we must make adjustments.

From then on our position at the ball could, from a distance, be mistaken for an impression of Quasimodo about to swing down from the lofty portals of Notre Dame. The uneven arms, the contorted legs, the sunken chest, the shoulder dropping horribly towards the knee, balanced only by the head with its one eye peering balefully down: only the great Parisian hunchback could assume such a posture.

Yet in spite of these contortions we do not expect to be disappointed when we make our first fretful lunge. We are victims of our own ingenuity. In much the same way as man has been able to set foot on the moon, so we have found ways to overcome physical shortcomings and strike the golf ball efficiently.

It only needs one perfect stroke from our misshapen forms for us to believe that there is plenty more where that came from. We may strike a thousand putrid shots but that single memory of perfection overcomes all the previous disappointments.

Of course, our cause is not assisted by the saints of the game who pass among us. We see them—the Hogans, the Palmers, the Nicklauses—equipped on similar lines but able to perform the necessary contortions. The truth is, though, that they are also playing the Game of Disappointment but on a higher level.

Disappointment is, therefore, relative to each player and acceptance of its inevitability is integral to the enjoyment of golf. So the next time you play, don't be disappointed because that's the name of the game.   CHRIS PLUMBRIDGE, *Sunday Telegraph*, 1992

❖

Not only is the stroke in golf an extremely difficult one, it is also an extremely complicated one, more especially the drive, in which its principles are accentuated. It is in fact a subtle combination of a swing and a hit; the 'hit' portion being deftly incorporated into the 'swing' portion just as the head of the club reaches the ball, yet without disturbing the regular rhythm of the motion. The whole body must turn on the pivot of the head of the right thigh-bone working in the cotyloidal cavity of the os innominatum or pelvic bone, the head, right knee, and right foot remaining fixed with the eyes riveted on the ball. In the upward swing, the vertebral column rotates upon the head of the right femur, the right knee being fixed; and as the club-head nears the ball, the fulcrum is rapidly changed from the right to the left hip, the spine now rotating on the left thigh-bone, the left knee being fixed; and the velocity is accelerated by the arms and wrists, in order to add the force of the muscles to the weight of the body, thus gaining the greatest impetus possible.   ARNOLD HAULTAIN, *The Mystery of Golf*, 1910

Neither your drive nor your approach nor your putt is determined by a psychic monitor flitting through your brain, nor by an unimaginable soul sitting in your glandula pinealis; they are determined by your nerves and muscles, by what you have been, by what you have learned, by what you have eaten and drunk. Never mind about psycho-physical parallelism. Keep muscle and nerve in good working order; be clean and honest; take lessons; learn; and be careful about what you eat and drink.   Ibid.

He has taken the game up with a passion since they joined the Flying Eagle, without getting much better at it, or at least without giving himself any happier impression of an absolute purity and power hidden within the coiling of his muscles than some lucky shots in those first casual games he played once did. It is like life itself in that its performance cannot be forced and its underlying principle shies from being permanently named. *Arms like ropes*, he tells himself sometimes, with considerable success, and then, when that goes bad, *Shift the weight*. Or, *Don't chicken-wing it*, or, *Keep the angle*, meaning the angle between club and arms when wrists are cocked. Sometimes he thinks it's all in the hands, and then in the shoulders, and even in the knees. When it's in the knees he can't control it. Basketball was somehow more instinctive. If you thought about merely walking down the street the way you think about golf you'd wind up falling off the curb. Yet a good straight drive or a soft chip stiff to the pin gives him the bliss that used to come thinking of women, imagining if only you and she were alone on some island.   JOHN UPDIKE, *Rabbit is Rich*, 1982

❖

The thing that astonishes me about golf is that 20 years after I first began to play it, I still have six separate and distinct methods of producing any given shot. The other thing that astonishes me about golf is that half of these separate and distinct methods are the precise opposite of the other three, and all of them work equally well. On—that is—their limited day.

I should have thought, after so long an apprenticeship, that one particular method would finally have risen to the top, and that with it, in the middle 80s, I should have been allowed quietly to play out my time. But no. As I stand on the first tee in the first round of, say, the Amateur Championship, at 8.30 a.m. with a nippy oblique breeze off the sea, I can strike the new ball in front of me in no fewer than six different ways, and that after a week of careful practice.

With this first shot there are three things I want to do. I want to get it off the ground. I want to be able to find it after I've hit it. And, rather specially, I want to avoid the one which shoots off the toe through the roof of the starter's tent. I might try the Hogan Crouch. I got the Hogan Crouch from looking at pictures of Ben Hogan, and with it drove the second green at Coombe Hill in March 1951. In the Crouch the knees are bent throughout the duration of the performance. The back is hollowed, and the buttocks (Hogan's word not mine) are thrust out in a sitting position. It certainly gives a feeling of exceptional power and mobility. But with the Crouch I must continue to crouch. If the knees should suddenly straighten, owing to centrifugal force, the clubhead passes over the ball, resulting in the clean miss (Wentworth, April, 1951; Stoke Poges, April, 1951; Walton Heath, April, 1951; Sunningdale, April, 1951, etc.).

The Upright Classical might be a wiser investment. I got the Upright Classical from looking at pictures of Henry Cotton, and with it nearly reached the ditch from the first tee on the first hole at Deal in a year I don't remember, because every Halford Hewitt, in retrospect, looks to me exactly alike. The difficulty with the Upright Classical is that my left leg, at impact, seems to have the rigidity and length of a telegraph pole, and a follow-through of more than a foot guarantees a permanent injury. This is the one which, incorrectly played, passes low over the heads of people not paying attention at cover point.

The Blacksmith's Convulsive has, of course, stood me in good stead before now, notably when I arose from the poker table at 9.10 one morning—previous best 6.30 p.m. to 8.15 a.m., Ballybunion, 1938—and slashed one right down the middle, five minutes later, in a West of Ireland championship at Rosses Point. The essence of the Blacksmith's Convulsive is to seize the club so tightly that the forearms become numb, and then let drive, disregarding pivot, weight-shift and any attempt to cock the wrists. The wrists, if cocked, snap off. The value of the Convulsive is that the stroke is always completed, one way or another, before the audience have had time to observe its finer points.

The Convulsive's companion, and, of course, direct opposite, is the *Sensitive*. A French word, pronounced *Sonsitive*. The thing about the *Sensitive* is that it relaxes me right down to the ground. The club is held so lightly in the fingers, and so marked is the absence of tension, that upon my word I'm in danger of dropping off to sleep. I discovered it one summer evening at Portmarnock, in 1936. We were fooling about on the practice ground, when I had occasion to tell one of my companions that he was trying to hit too hard. 'Just drop the clubhead on the ball', I said, 'like this.' I dropped the head of a No. 5 on to the ball and it took off like a bullet, to crash full pitch through the secretary's window—as we paced it out later—207 yards away. I could certainly use it now, except that with the *Sensitive* the backswing gives me the impression of going on for ever. I have a tendency to lose my nerve, and to wonder if the ball will still be there when, like General MacArthur, I return. Under these circumstances I'm inclined to lash out suddenly from the level of the waist, and off she goes in that special kind of hook so quick that the ball becomes egg-shaped in its efforts to take the bend.

Round about now, on the first tee in the opening round of the Amateur Championship, I often begin to wonder which hand goes on top of the club, in the grip which I am accustomed to use. But this is only a passing phase. I have more serious things to consider. Whether, for instance, at this eleventh hour, to use the Utterly Inside Out, or the Retired Colonel's Up and Down. I once played the Utterly Inside Out by accident, into the heart of the green at Calamity Corner, Portrush. There was a gale blowing from the sea. To keep the clubhead as close to the ground as possible, and thereby out of the worst of the weather, I drew it back stealthily in a half circle round my legs. I had no expectation that anything very much was going to happen, but without warning, and for the first time in my life, I felt that if we were to be fortunate some time in the future to return to the target area I was going to have a solid left side to hit against. I could actually see my left shoulder, as large as life beneath my chin. An astounding spectacle, in view of the fact that I'd never been privileged to get more than a glimpse of the elbow before. Round we came. And off he went—actually boring in from the left, to pitch downwind, check, and finish three feet from the hole. At the next, I tried it again, with the driver. We gave up looking for it almost immediately. Another couple was pressing behind. But the caddy, in that rather charming Belfast accent, said that never before in his life had he seen a shot which had travelled 270 yards without actually crossing the front of the tee.

With only a few seconds left before my name is called I decide it will have to be the Retired Colonel's Up and Down. This is the simplest style at my command. I stand in front of the ball and hold the club just firmly enough to stop it falling out of my hand—a semi-*Sensitive*—and then I pick it up and move it back and when I think I've got it back far enough I move it forward again, and if the ball, however briefly, manages to get itself tangled in the lot I do swear I'm well pleased.

But the Retired Colonel's Up and Down got me drummed out of the preliminary trials for the Walker Cup. From the first tee at Temple I knicked one so finely with the Retired Colonel, off the toe, that it would have carried away the starter's tent and the starter, if the starter and his tent had been there. As it was, it nearly felled three members of the selection committee. Discussing the technicalities of the thing afterwards, Mr Raymond Oppenheimer told me that it had seemed to him and his fellow selectors that my best prospect of making contact with the ball, as from the top of the swing, was to strike at it vertically, downwards, keeping the head well back and out of harm's way.

But now it is my turn. The die has been cast. The name has been called. After two attempts I tee up a new ball. I initiate the backswing with a slight forward press—and suddenly every one of my six systems becomes fused! From the Hogan Crouch, with a grunt, into the Upright Classical! *Sensitive* at the top of the swing! The club is lying on my collarbone! Quick—the Blacksmith's Convulsive—take hold! We're coming down. Ease her in. More. A little more. The Utterly Inside Out. Too much! I'm going to strike my hip! Quick! The Retired Colonel! The Up and Down! Level her off! I've got it! Contact! She's away! A low snakey thing, scorching the rough all along the lefthand side of the fairway. And all of 150 yards. Well, by God, that's one I haven't seen before.

PATRICK CAMPBELL, 1951

LEARNING THE GAME

Sam Snead's first strokes were with a swamp maple branch carved into the form of a rough club, and he learnt by caddying as a tiny lad, barefoot even in the frost. Francis Ouimet may not have suffered the same degree of hardship, but his story is of a similar youthful passion for the game, overcoming the obstacles of lack of equipment and access to a course.

Long before I ever had a club I had golf balls enough to last me for years. But the balls without a club were not very useful. Golf was so new to America in 1900 that it was difficult to get clubs. They never got lost, and were rarely discarded. The balls, however, seemed to have plenty of life in them, their varied markings held some sort of fascination for me, and it was fun watching them bound from rocks and other solid substances.

After I had hoarded golf balls enthusiastically for two years, someone gave my brother Wilfred a club. When Wilfred was busy caddying, I helped myself to that club and used it to knock some of my hoard round the backyard. I was careful to put Wilfred's club back in its place before he put in an appearance. Otherwise, I felt, there might have been a family riot. Occasionally a tournament was held at the Country Club, and on those days and after school I would stand on the edge of a fairway and watch the golfers go by. If I saw someone play an exceptional stroke, I watched how he did it and hastened home to take Wilfred's club and set about trying to put into practice what I had seen. Those efforts must

[45]

have been funny, but they were, after all, the beginnings of my game, such as it is.

I can remember vividly the first Haskell ball I ever found. It was in the autumn of 1902, and I was nine years old. Wilfred was a caddie boy at the Country Club, and the ladies were having their National Championship. On the way home from school I picked up a nice new ball. It was unlike any other I had ever seen and seemed much livelier. I showed it to Wilfred and he told me it was one of the new rubber-cored balls. Few had them, and Big Brother tried his best to talk me into parting with it. Nothing doing. I played with it, bounced it, and used it until the paint wore off. I got some white paint and painted it. Mother was baking some bread in a hot oven and I sneaked my repainted Haskell into the oven, thinking the heat would dry the ball.

Mother smelled something burning and went all through the house trying to discover the cause. She found nothing, but the odour was so strong, and she was so worried that the house was burning up that she kept on searching. Finally she opened the oven door, and the most awful smell in the world came out of the newly made batch of bread. It was ruined—and so was my prize, the Haskell. The heat had melted the gutta-percha shell and there was nothing left of the thing but a shrivelled-up mass of elastic bands. I learned then and there how Doctor Haskell made his golf balls and why it was that the rubber-cored ball was vastly superior to the solid gutta.

The Haskell crowded the gutta off the courses and made the game much more enjoyable to play. At any rate, I could play the rubber-cored better than the hard ones, and my interest in the game increased. Behind our house was a cow pasture, and here Wilfred, with the mind of a golf architect, built three holes. The first was about a hundred and fifty yards long, with a carry over a brook. The brook was a hundred yards or so from where we drove. When he hit a shot well, Wilfred could drive close to the green, but it was far beyond my reach. As a matter of fact, the very best I could do was to drive into the brook. The second hole was very short, hardly more than fifty yards. The last was a combination of the first two, and brought the player back to the starting-point. We used tomato cans for hole rims. As I visualize that old course of ours, it was the most difficult one I have ever played, because it contained a gravel pit, swamps, brooks, and patches of long grass.

We—or rather Wilfred—had selected only the high and dry pieces of land, which were few and far between, to play over. A shot that travelled three yards off line meant a lost ball, and it was well we had plenty!

Wilfred made trips to Boston from time to time and discovered that Wright and Ditson had a golf department with a man named Alex Findlay in charge. He discovered also that a good club could be got in exchange for used golf balls, and that three dozen would be a fair exchange for the best club made. From one of these visits Wilfred brought me home a mashie, and for the first time in my young life I was independent so far as playing golf was concerned. I had my own club, balls, and a place to play. What more could anyone ask?

A lawn-mower kept two of the greens in fair condition, but the one near our house was used so much it was worn bare and had no grass whatsoever on it. You see, while we were waiting for a meal we fiddled round the hole and the grass never had a chance to grow. One advantage, from Mother's point of view, was that she always knew where to look, and it was a simple matter for her to call us into the house. We fooled round that particular spot early in the morning and long after dark, and it was small wonder that my interest in golf increased because, with all this practice, it was natural enough that I should notice some improvement in my play. Mother thought I had gone crazy, because golf was the only thing I seemed interested in.

I had more time to devote to the game than Wilfred. He had chores to do round the house and barn, and, being older, he was the one called upon to go on errands. They say practice makes perfect, and I believe it. After striving for weeks and months to hit a ball over the brook, and losing many, I finally succeeded. A solid year of practice had enabled me to drive accurately, if not far, and one Saturday morning, after trying for an hour, I drove a ball as clean as a whistle beyond the brook.

When I told Wilfred of my accomplishment, he received my story with a good deal of doubt. I had now acquired a brassie to go with the mashie, and I invited my brother out to the pasture to see what I could do with it. Whether I was tired out from my earlier efforts or not I do not know, but I failed utterly, and Wilfred naturally was more sceptical than ever. The next day was Sunday, and after I returned from Sunday School I went at it again. This time Wilfred was with me, and I definitely convinced him by hitting two balls out of three over the brook. It soon got to be a habit, and I was quite disgusted with myself when I failed.

A good many tournaments were held at the Country Club and the best golfers gathered to play in them. Soon I was old enough to caddie, and as a youngster of eleven I saw in action such great golfers as Arthur Lockwood, Chandler Egan, Fred Herreshoff, Jerry Travers, and Walter J. Travis among the amateurs, and Alex Campbell, the Country Club professional, Alex Smith, Tom McNamara, Willie Anderson, and many of the prominent professional players. If I noticed anything particularly successful in the play of any of these golfers, I made a mental note of it, and when opportunity afforded I set out to my private course and practised the things I had noted.

Therefore, you see, I was brought up in a golfing environment and learned to love the game. I read in magazines or newspapers anything I could find relating to golf, got a few of the boys in the neighbourhood interested in the game, and jumped into it head over heels. One day I caddied for a dear old gentleman named Samuel Carr. Mr Carr was a golfing enthusiast, and, furthermore, always most considerate of the boy who carried his clubs. All the boys liked him. Playing the eighteenth or last hole one day, he asked me if I played golf. I told him I did.

He asked me if I had any clubs. I replied that I had two, a brassie and mashie.

'When we finish, I wish you would come to the locker room with me; I may have a few clubs for you,' he said.

I took Mr Carr's clubs downstairs to the caddie shop and hustled back. He came out with four clubs under his arm, a driver with a leather face, a lofter, a midiron, and a putter. I think it was the biggest thrill I had ever got up to that time.

Early mornings—and when I say early I mean round four-thirty or five o'clock—I abandoned my own course and played a few holes on that of the Country Club, until a green-keeper drove me away. Rainy days, when I was sure no one would be round, I would do the same thing. Complaints concerning my activities arrived home, and Mother warned me to keep off the course, usually ending her reprimand by saying that the game of golf was bound to get me into trouble.

I was so wrapped up in the game, however, I just couldn't let it alone. One summer, tired of my own layout, I talked a companion,

Frank Mahan, into going to Franklin Park with me. Franklin Park was a public course and we could go there and play unmolested. We set out one Saturday morning. To get to Franklin Park, we had to walk a mile and a half with our clubs to the tramline. Then we rode to Brookline Village, transferred there to a Roxbury Crossing tramcar, arrived at Roxbury Crossing, and changed again to a Franklin Park tram. After getting out of the last tramcar, we walked about three quarters of a mile to the club-house, left our coats— that is all we had to leave—and then six full rounds of the nine holes, a total of fifty-four holes.

Then we went home the way we had come, completely exhausted. All this at the age of thirteen!

Another thing I like to remember is the day I was selected by Dan McNamara, the caddie master at the Country Club, to act as caddie for a gentleman named Theodore Hastings. Mr Hastings was peculiar about his golf: he invariably played alone. As we walked toward the first tee, he asked me if I played. I told him I did. He asked me where I lived. I told him. He said, 'Get your clubs and we will play a round.'

Of course, caddies were not permitted to play on the course at all, but when Mr Hastings invited me I forgot all about regulations, dashed home for my set, and all running records were broken in getting back to that first tee.

The Country Club course was not as difficult then as now, but for all that it was one of the leading courses in the country and was hard enough for anyone. I played the first nine holes in thirty-nine strokes, and Mr Hastings was considerably impressed. I not only lugged my own clubs, but his as well, and was having a marvellous time. The fifteenth hole passed directly by the caddie house, and Dan McNamara usually sat in a chair overlooking the caddie shed and with a perfect view of the fifteenth fairway. Furthermore, he was a disciplinarian. I have since learned that a good caddie master has to be.

My play continued satisfactorily through the fourteenth, but then I began to think of Dan! I hit a good tee shot over the hill, and, walking round, I cast a glance toward the caddie shed. There was Dan. I doubt if I have ever been as nervous before or since. I topped my second shot and missed the third. It was a simple hole to make in five. I put my fourth into a trap and needed three to get out. The hole cost me a ten.

Once out of Dan's sight, I steadied a bit and finished the round with a score of 84. Delighted beyond words, I was ready to meet Dan in all his fury. Mr Hastings came to my assistance, though, signed my caddie check, and told Dan he had had a most enjoyable afternoon. Dan was reasonable, too, seemed interested to know what score I made, and wanted to know also what happened to me on the fifteenth. I truthfully told him I expected to see him come running after me, and was just frightened to death.

FRANCIS OUIMET, *A Game of Golf*, 1933

A gentler, more sheltered version of the same story of boyish enthusiasm.

When I began to play it, golf was not considered a suitable game for the young. It induced a spirit of selfishness, it was not vigorous enough, it was regarded as not quite 'healthy'. My mother I know was warned of the perilous influence it might be upon my malleable four-and-a-half-year-old character. But here we were opposite the Heath and the doctor had said I must be out in the air as much as possible. Why not, said Mr Padwick, give him a golf club? This solution had not occurred to Mama; she knew nothing about golf. She consulted friends: the general opinion was that I was too small to be affected morally and that it could do no harm, particularly since I was not taking to golf deliberately, as a game, but simply for the sake of my health. To keep this therapeutic approach pure, it was felt that I should not have lessons. Anyway I was too small. Nor was expensive equipment involved. Mr Padwick was going to give me a cut-down club and an old ball or so, and after that I could supply myself with the lost balls which I would find in the gorse-bushes. Going out ball-hunting became for me a delicious kind of adventure in itself. I spent hours tunnelling in the gorse worming about in the musty dry undergrowth or poking in the mud of the stream waiting for a bite. I could tell the feel of a ball from a stone in an instant. There was a shabby old man with a mongrel who spent every day, all day long, ball-hunting. 'Old Pocock' never gave me a ball once. He was a bit gone in the head, my mother said, but quite harmless. He hoarded golf balls like a demented squirrel; his rooms must have been piled with them. He had his 'rounds', which I soon learned, for it was no use following after him. His eyes and nose were always damp and when we met I'd say, 'had any luck?' He shook his head like an out-of-work mute. You could see the pockets of his old blue overcoat, which he wore summer and winter, knobby with spoil.

The club Mr Padwick gave me was old. He had cut down the hickory shaft to about two feet long. The head was iron and chocolate-coloured, for the outer rust had almost a patina. He said it was a mashie. I loved it at once.

My mother's attitude towards golf was wary. I am sure that she never really wanted to play herself. When she found she had to, she played with a sort of resigned dissociation. But at first all was well. She came out with me once or twice and sat in the shelter of a gorse bush while I flailed away. She could not help me, and beyond telling me to watch other players and especially Mr Edmunds, the 'pro', left me alone. Soon, the cure was obviously working. Golf was making me sturdy.

I hated to be watched. Once through the gate I trotted or ran, bearing left round the slope until I was out of eyeshot. Then I'd drop a ball wherever I was and hit at it. Mr Padwick had told me a mashie was a club for getting 'out of the rough'. Even on the golf-course itself there was little difference between rough and fairway and I did not begin to venture on the real course. I kept to edges and out-of-the-way stretches. If I managed a good shot—say ten yards—I ran as hard as I could after the ball for fear I should lose it. I often did, even so, and padded up and down round and round with my eyes on the ground till I pounced on it. Then I'd hit it again.

This must sound dull. But I was intensely happy. I loved the wiry shiny knots of coarse grass, all the green and yellow and dun colours and the way one's boots could almost slide over it. I felt the whole scene, right to the Downs, close round me as if it were something I wore hidden against my naked skin. It clothed me secretly. Every shot I made was exciting. It was gorgeous to see the ball up in the air. But I often topped it and it would hop and dribble miserably among the tussocks. I came after it hot and stubborn, and without giving myself time to settle would slash at it again. Sometimes these angry slashes were tremendous successes. I learnt self-control instinctively and naturally.

Soon I was imposing tests on myself. I would drop three or four balls in a particular place from which I must clear thick rough and land on safe ground. If any of my shots failed I had to repress my yearning to go after it at once; I had to 'mark' the place in my head and then play the next shot which might succeed or fall into some different quarter of the slough. Total success was elating; but total failure, so to speak, even more if one successfully found each ball—perfect if one walked straight on them. I think I learnt a certain tenacity of purpose from this game.

PATRIC DICKINSON, The Good Minute, 1965

As I grew so I haunted the golf pro. I trailed after him when he played; when he practised I was his retriever. I lay doggo, my eyes never left him. Three dozen shots, and I was off with the linen bag whirling behind me full of air—it felt lovely. Then there in an area so confined that I was always awestruck lay the balls like a crop of mushrooms: the Colonels, the Silver Kings, with their bramble-and-dimple markings and the new square lattice. I followed Mr Edmunds to the 'pro's shop'. I loved to see him planing the hickory shafts and chiselling the persimmon heads for drivers and brassies. I learnt all the clubs and their wild names: cleek, niblick, lofter, putter,—mashie I knew—baffle, mid-iron. 'Iron' clubs were not rustless. I spent hours polishing for Mr Edmunds. I loved it. At first a strip of emery paper was stiff and graty. Soon it became malleable, finally a rag. Doing the hose of an iron was gorgeous. You stood the club upright—upside down with its head in the air, wrapped the paper round the hose, held it loosely and whirled the head round and round with the other hand. This was not so easy as it sounds since I was very little taller than the club. I worshipped Mr Edmunds. I always called him 'Mr Edmunds' and he called me 'Master Pat'. He called my mother 'Ma'am' and she gave him a 'Christmas Box' for Mrs Edmunds. He was tall and freckly fair; he had a moustache which I admired. I did not know then, but seeing it in my mind's eye I am sure he wore it to be as much like the great champion James Braid as a fair man could. I do not know what he told me about them, but long before I was eight Harry Vardon, Sandy Herd, J. H. Taylor and James Braid were 'huge and mighty forms that did not live like living men' and certainly they moved through my mind in a strange procession.

From Mr Edmunds I absorbed golf physically, mentally and spiritually. I knew him to be a kind gentle man. My mother entirely approved of him. One day he gave me a wooden club he had made especially for me. It had a large head, with a brass sole to polish and about the loft of a modern No 5 Wood, but it was my brassie.

I scrounged a putter and got it cut down. I reckoned now that I had a set of clubs. All I needed was a bag. My mother very sensibly wrote to Hamleys. As a surprise, I was given the result. I was not at all grateful, I was furious. It was not a proper bag at all, it was a toy bag. I had to be persuaded to use it as a stop-gap until I—and my clubs—grew a little.

My chagrin passed quickly. Mama must have talked to the secretary, for quite soon after the arrival of my bag she told me I was to be allowed to play on the course so long as I got out of anybody's way. This was in theory bliss. In practice it was not. Between each tee, and fairways which were not greatly distinguishable from it, stretched The Rough. I was too small to carry it from the tee; too proud to skip it and start from the fairway. If golf was to be played it must be played properly and correctly. Alas, only too conscious of his doom the little victim played. This passage of the Rough of Despond may have been good for my character. I realise that it was bad for my golf. My driving has always been my weak spot. I am certain that this is because I knew every time I stood on the first tee at Petersfield, aged six, that however good a drive I hit I would land in The Rough. I knew it as I put down my bag and took out my brassie. I knew it as I went to the tee box and took out enough sand to build my tee. I knew it as I teed up my ball—and knew worse: I should probably lose it. This made me apt to lift my head too quickly to see where it was going. Golfers know what ensues; non-golfers may guess. But give me a good drive: I would be after it, fast grabbing my clubs on the way. Got it! Out with the mashie. Swat! say, a yard gained. Swat! three yards half-right. Swat! Swat! Six yards, and it never bounced up at all. After it in a flash, leaving bag behind. Up and down up and down in a close area. It must be here. Sometimes I never found it. Sometimes leaving my mashie to mark where it lay I would go back for my bag, return, and in frenzy find I had lost both club and ball. Once I ran all the way home to get Mama to help. It must have been fascinating to watch my slow rushes and frantic slashes. As I approached the further edge of The Rough having scored about ten there was a final peril. There were two jutting bastions of gorse and across the strait between them slimed a filthy stream whose maw was full of rotting balls. Sometimes my final hack would rise hopefully, fly and fall like Icarus to disaster: or hitting extra hard I would slice or pull into the gorse. I searched and searched. If I found my ball I dropped it on the other side. If I did not I penalised myself two more strokes and dropped another. And as I climbed the rising fairway I looked back to see Pocock slide from some hideout in the gorse, a dark remorseless ogre, and begin to probe.

The play from the ditch to the green was truly child's play. Many golfers dread putting, even the greatest. I have always loved putting and found it easy and immensely enjoyable. Equally I have always loved approach-play of all kinds. The first hole at Petersfield is an allegory of my golf. I was ridiculously obstinate about this hole. Sometimes I deliberately dared the Fates by putting down my best ball. I do not think I have learned yet that they never grow tired of winning.

After a year there came a wonderful day when I ran shouting home, down over the slope from the last green, hopping and leaping through the bushes, getting myself and my clubs jammed in the iron cuckoo-wicket, dashing across the road, shouting for Mama. She shot out of the front door just as my bag of clubs got between my legs and I crashed most dramatically at her feet.

'What is it, darling?'

'I've broken a hundred! I've broken a hundred!'

She considered the statement. I was not injured. I had broken a hundred. What was that? By this time I had untangled myself and produced a crumpled scorecard. She had no idea that this was a moment of great triumph. She said 'Are you sure you've added it up right?' I had counted every shot. Round after round some disaster overtook me and I realised at about the seventh that I had too few left to make my ambition possible. That morning I had stood on the ninth tee, my heart punching my ribs. I could do it. Three to clear the rough, three more to get up the hill and on to the green and two putts. Mr Edmunds had told me—it was the only direct advice I remember he ever gave me—that no good golfer ever took more than two putts. This was Law, the greens were small, and I seldom did. And eight would give me a ninety-eight. I hit a beauty from the tee, almost half-way across the rough. The ball lay well and my noble mashie took over—I was clear in two. Another and I was half-way up the hill. Was I on the path? This was the last appalling pitfall because the path was rutted and full of loose silvery sand. I was on the path. There are bitter self-pitiful moments in later golfing life when the victim swears, 'This would happen to me.' The Fates make a note and see that it does again, soon, and worse. I was still too young for this treatment. They simply asked me to play from the path again because they hadn't my first try. My third try got out. I had played six. But now I lay easy on grass and played over the crest to where I could see the top of the flag. When I came up, my ball was about three yards from the hole. Two putts made ninety-nine. I felt I was now really a golfer.   Ibid.

Soon after I had got a scholarship, in May 1928, Mr Jeston asked me if I would like a round at Sandwich. I had all my six clubs at school by then in a proper bag in which Mama had sewn a tuck, like there were in my shirtsleeves.

We set off in his Armstrong-Siddeley immediately after lunch. The sun was hot. As we walked to the first tee of the first 'Championship' links I had ever seen, I went sedately, for I had a *real* caddy. Mr Jeston bought me a new ball—a Dunlop Maxfli—at Whiting's shop. I had a card too, I was going to mark my score. Ahead was sheer emerald of fairway, grey-green of rough and sand-hill. Above, the sky was starry with singing larks; there was a tiny shimmering breeze blowing into my face from Pegwell Bay. It made the whole blue sky seem thin and weightless, as if it were a flag. I know this is a curious simile, but that is what I felt then, so I can put it no other way. I was brimful of joy, or I thought I was till the old caddy took out my driver and said, 'Let's see you swing'. Then he turned to Mr Jeston. 'He's a real golfer, then.' He pre-pared to make my tee. No. I knew best how I liked it. I felt his pleasure; he was for me. I was brimful of joy as I took my stance for my first drive. I played from the most forward tees. I could carry the rough and did. Three hours later I walked off the eigh-teenth green. To describe what had happened is impossible. I know that Mr Jeston, and I think that my caddy, were happy-sad for me because they knew that this could never happen to me again. I did not know. Not even the lightest frame of consideration enclosed me. I was in a dream come true before I had dreamed it. I had been, as it were, in a controlled ecstasy, not aware that ecstasy existed, only a little aware that control had come from outside me.

Let me put down the traces of this divine visitation; they have a perfect, therefore once flawed, beauty. There are eighteen holes on a golf links, at none of which even the greatest player is expected to take more than five strokes. At some he should take four, at 'short holes' three. I did seventeen of the holes in five strokes each; one I did in four. Eighty-nine strokes. My unavoidable failures were the short holes which presented difficulties I was physically unable to overcome. My fives at the bogey-fours were never less than commendable, my fives at the bogey-fives were exciting, sometimes magical triumphs. All the round long 'it was shining, it was lovely'. I was so light with joy I could feel the grasses under my feet growing me taller, as they grew; I could feel through my shoes the change to the solid sheeny putting greens which somehow sounded hollow, if you dropped a ball on to them. (To me, they still do.) There were tiny flowers and traces of rabbits and the scent of hot dry sand, dry sweet grass, and the bitter-sweet of salt—a tasty outdoor kitcheny sort of smell. I remember that my caddy was tensed and came to the 'Canal' hole with me stiffly like a bent walking tree. It is the one time I consciously see him. I got my five for him. My only four came at the last short hole, the sixteenth. Then six full wooden-club shots and four putts saw me home.

I have often tried to write a poem about golf, but I have never been able to. This round *was* my poem of golf; perhaps that is the reason.   Ibid.

I played a lot with Major M—— when I was new to the game, at the age of eighteen, for the reason that he was the only member of the club who had nothing to do from Monday to Friday either.

Being at that time a novice, I believed that anyone's play could be improved by a careful study of a book of instructions, and was working on this particular day on the full arc of the back swing, paying special attention to the need for a firm grip with the left hand at the top. I decided to pass on the message to the Major, who seemed to be in as urgent need of it as any man I'd ever met.

His method of striking was to bounce the shaft of the club on his right shoulder so that the down swing was initiated by a ricochet over which he had no control whatever, so that the ball could travel in any direction but never more than two feet above the ground.

The show-down came when he'd been bouncing the shaft with extra ferocity in an effort to hack a ball called a Goblin out of the pervading mud.

We played on a 9-hole course which consisted of three flat fields put together and divided by threadbare hedges, a venue offering little of interest to the power player except that it was downhill on a bicycle from where I lived. On this Tuesday afternoon, with an autumnal fog already settling in, the Major and I were naturally the only ones out.

It was at the short eighth that I asked him bluntly about the bouncing, an idiosyncrasy which had been weighing on my mind for some weeks.

It seemed a good time to deal with it, in view of the small, disused quarry on the right of the tee—the only feature of interest on the whole course.

We'd spent many hours in there already, poking about in nettles, brambles and long wet grass for the Major's tee-shots and not infrequently, I had to allow, for my own as well, and though we'd inevitably uprooted or flattened a good deal of this undergrowth plenty of holding stuff still remained. Furthermore, I knew the Major was down to his last Goblin, and I had no desire to lend him one of mine.

'Excuse me, sir,' I said, 'it's only a suggestion, but have you ever tried getting your hands a little higher at the top of the swing?'

Up till then I'd only criticized the style of my contemporaries and so was not prepared for the reaction of an older man. The Major glared at me with a malevolence which was surprising, in view of my innocent desire to help. 'You play it your way, sonnie,' he snapped, 'and I'll play it mine.'

He bent down, to tee up his Goblin. He was, in fact, three-up, owing to a stiff-wristed putting method I was trying out for the first time.

The foliage in the quarry looked, however, particularly dank and uninviting, so I tried again. 'But,' I said, 'you let your grip go altogether at the top of the swing so that the shaft bounces off your shoulder and ruins your arc.'

'Arc?' said the Major. 'Arc? Let me tell you something,' he said. 'Jimmy Braid always slackened his grip at the top of the swing and he'd have seen any of you young artichokes off any day of the week, wet or fine.'

In later years I learnt, of course, that anyone over the age of fifty cannot be shaken in defence of his own swing, despite the fact that he's never played a medal round in less than ninety-three. At eighteen, however, I still believed that even the mature player would be interested in advice. 'I know, sir,' I said, 'but I'm sure if you tried holding the club just a little more firmly at the top it would give you a lot more length off the tee.'

'Ha!' the Major said, dismissing this sound principle out of hand. He squared up to his Goblin. He started his back-swing or, rather, initiated the jerk that normally lifted his mashie into the air, and all at once something looked different. His knuckles were shining white, indicating a grip on the club liable to squeeze the plug out of the shaft, if he were able to maintain it.

It was a phenomenon which was later to become only too familiar—that of the player who contemptuously rejects all advice on the grounds that it's drivel and then furtively tries to apply it, in case it might work—but I was too alarmed for the Major to consider it now.

He had reached the top of his swing—or whatever it was—and was standing on tip-toe with his left arm across his eyes and the club raised vertically in the air, presenting the appearance of a monk in plus-fours and a check cap about to haul down on the bell-rope for the opening chimes of the Angelus. Under the circumstances, it seemed improbable that he intended to continue with the stroke.

I was about to step forward and break the deadlock by taking the club away from him when he slashed suddenly and viciously downwards, making an attempt at the same time to turn his hips to the left, trying to deflect the club-head from the vertical into the horizontal plane.

The result was something I'd never witnessed before, and have never been privileged to see again.

The head of the club buried itself in the mud nearly a yard behind the ball and remained there, leaving the shaft standing upright like a flagstick. It was not attended by the Major. The force of the impact had torn it from his hands. With nothing to hold on to he swung round, staggered forward and trod fairly and squarely on his Goblin, obliterating it from view. He stood there, facing the hole, vibrating a little and peering about to see where the ball had gone. Behind him, the mashie slowly keeled over and fell to the ground.

I picked it up, wiped some of the deposits off the head, and held it out to him. 'That was bad luck, sir,' I said, for the sake of saying something. 'I think your foot slipped a little. Have it again.'

He took the club from me, with a set face. He examined it briefly, as though to make sure it was his own. 'All right, then,' he said, 'where's the ball?'

I had to hack it out of the ground with my own club. I teed it up for him again.

'If you wouldn't mind minding your own dam' business,' the Major said, waggled once, bounced the shaft on his shoulder and slashed the ball straight into the quarry. I followed him immediately afterwards, out of nervousness. After poking about unsuccessfully for some time we walked in, the Major having run out of ammunition.

On the steps of the clubhouse he spoke for the first time since the quarry. 'Interfering with a chap's game,' he said, 'is dam' nearly cheating, and I don't like it.' He allowed the door to swing back in my face. PATRICK CAMPBELL, *How to Become a Scratch Golfer*, 1963

PUTTING

Everyone is agreed that putting is a game within a game. Hogan, late on in his career when he was not putting well, is said to have declared that it should be done away with altogether and that the hole should be awarded to the player getting nearest to the pin. Certainly it is the department of the game where the consistent skill of the leading professionals most clearly sets them apart from the amateur.

The idea of golf is to get the ball in the hole. Since the green is where the hole is, that's the place where emotion is ultimately released, or expressed. When a player holes a crucial short putt, or rolls one in from sixty feet, it is on the green—on the 'dance floor,' as golfers like to say—where he punches the air and does a little jig of celebration.

But the metaphor of the green as a 'dance floor' has other connotations.

The green is a wonderfully smooth and uncluttered place compared to the rest of the golf course. There are no trees or sand traps, no shin-high grass, no ponds or streams to impede the journey to the place where the hole is. Once on the green, the golfer is out of harm's way, more or less, for while putting, which is the business done on the green, has its own set of perils, they are much more subtle than all the others he faces in getting there. While making his way to the green the golfer is melded with, if not swallowed up by, the trees and sand traps, ponds and streams. He is but part of a rather 'busy' landscape—and one of the smaller parts at that. But when he steps onto that broad, clean expanse of tightly mown grass that is the green, he is out of the clutter and stands out in sharp relief.

Every golfer knows the sensation of singularity that comes from walking onto the 'dance floor.' It has something to do with ego. And the feeling is magnified for competitive golfers on the big-time tournament circuits. When they walk through the gallery onto the green, they might be likened to ballroom dandies out to waltz a pretty lady while those who are perhaps less brave—or lucky— watch from the balcony.

AL BARKOW, Gettin' to the Dance Floor, 1986

While, on the whole, playing through the green is the part of the game most trying to the temper, putting is that most trying to the nerves. There is always hope that a bad drive may be redeemed by a fine approach shot, or that a 'foozle' with the brassy may be balanced by some brilliant performance with the iron. But when the stage of putting-out has been reached, no further illusions are possible—no place for repentance remains: to succeed in such a case is to win the hole; to fail, is to lose it. Moreover, it constantly happens that the decisive stroke has to be made precisely at a distance from the hole such that, while success is neither certain nor glorious, failure is not only disastrous but ignominious. A putt of a club's length which is to determine not merely the hole but the match will try the calmness even of an experienced performer, and many there are who have played golf all their lives whose pulse beats quicker when they have to play the stroke. No slave ever scanned the expression of a tyrannical master with half the miserable anxiety with which the performer surveys the ground over which the hole is to be approached. He looks at the hole from the ball, and he looks at the ball from the hole. No blade of grass, no scarcely perceptible inclination of the surface, escapes his critical inspection. He puts off the decisive moment as long, and perhaps longer, than he decently can. If he be a man who dreads responsibility, he asks the advice of his caddie, of his partner, and of his partner's caddie, so that the particular method in which he proposes to approach the hole represents not so much his own individual policy as the policy of a Cabinet. At last the stroke is made, and immediately all tongues are loosened. The slowly advancing ball is addressed in tones of menace or entreaty by the surrounding players. It is requested to go on or stop; to turn this way or that, as the respective interests of each party require. Nor is there anything more entertaining than seeing half a dozen faces bending over this little bit of moving gutta-percha which so remorselessly obeys the laws of dynamics, and pouring out on it threatenings and supplications not to be surpassed in apparent fervour by the devotions of any fetish worshippers in existence.

A. J. BALFOUR, *Golf*, 1890

❖

When a putter is waiting his turn to hole-out a putt of one or two feet in length, on which the match hangs at the last hole, it is of vital importance that he think of nothing. At this supreme moment he ought studiously to fill his mind with vacancy. He must not even allow himself the consolations of religion. He must not prepare himself to accept the gloomy face of his partner and the derisive delight of his adversaries with Christian resignation should he miss. He must not think that it is a putt he would not dream of missing at the beginning of the match, or, worse still, that he missed one like it in the middle. He ought to wait calm and stupid till it is his turn to play, wave back the inevitable boy who is sure to be standing behind his arm, and putt as I have told him how—neither with undue haste nor with exaggerated care. When the ball is down and the putter handed to the caddy, it is not well to say, 'I couldn't have missed it'. Silence is best. The pallid cheek and trembling lip belie such braggadocio.

WALTER SIMPSON, *The Art of Golf*, 1887

❖

THE DOUBTER'S FATE
This is a very simple job,
And when I have holed the ball,
I shall be certain of my half-crown,
Still, I must be careful. It is very easy to miss these short putts;
And I have missed many thousands, costing me
Many pounds—scores of pounds.

And now that I am up against it,
And looking at this putt,
It does not seem quite so easy as it did at first.
It will require most careful management—a most delicate tap,
And very accurate gauging of strength.
One needs to be very cool and deliberate over these things.
One's nerves, and stomach, and liver must be in prime condition.
I wish I had not been out to dinner last night.
Was it Willie Park or Ben Sayers
Who said that the man who could putt could beat anybody?
I believe him—Willie or Ben.
This is really a most awkward putt.
The green looks slower than the others. It is very rough.
Why don't the committee sack the greenkeeper,
Who ought to be a market gardener?
It is like a bunker between
My ball and the hole. Such very rough stuff.
One, two, three—six—nine—why?
There are eleven big blades of grass
Sticking up like the rushes at Westward Ho!
The grass becomes so very stiff and wiry in this very hot weather.
(Yes, it is too hot to putt properly).
My ball will never break through this grass.
It is one of the hardest putts I have ever seen.
I wish I had more loft on my putter.
I was an ass not to bring that other one out from my locker,
Where it is eating its head off (so to speak).
I think, also, that a little cut would do this putt a lot of good.
But how? The green slopes from the left;
Yet it seemed to slope from the right.
Also, it goes downwards to the hole.
This is a perfect devil of a putt!
I know my stance for putting is not good,
But Harry Vardon says that every man has his own stance,
So perhaps it is all right.
But I had better move my left foot; it seems in the way.
I see that two—four—six—seven of the pimples on his ball
Are quite flat.
Nobody can putt with a ball like that.
A man ought to be allowed to change his ball
Even on the green at all times like this,
I must allow for those pimples.
Confound that fellow Brown!

He seems to be waiting.
And he is smoking his dirty shag so much
That I can hardly see the hole for smoke.
If I lose this hole I shall lose the match.
I am quite with Johnny Low in his new idea for handicapping,
When he says some of us should be allowed to play
Our bad shots over again.
In that case I would have one good smack at this ball
To get the strength and the hang of
Everything. And I am certain—yes, I am quite absolutely certain—
That I would hole the ball next time.
However, what does it matter?
Better men then I have missed such putts,
And I am not a chicken—live a hard life—lot of work—
Office to-night—awful day to-morrow.
And as the wife was saying—
Let me see. Oh! hang this putt!
He can have his half-crown if he wants it,
But I am going to have one good smack
At this ball. Now
No, that was wrong. Now, yes, yes—

•   •   •

My godfathers!
And my godmothers!
I have missed that putt again!

HENRY LEACH, *The Spirit of the Links*, 1907

❖

Do not get into the habit of pointing out the peculiarly salient blade of grass which you imagine to have been the cause of your failing to hole your putt. You may sometimes find your adversary, who has successfully holed his, irritatingly shortsighted on these occasions. Moreover, the opinion of a man who has just missed his putt, about the state of the green, is usually accepted with some reserve. HORACE HUTCHINSON, *Hints on Golf*, 1896

GAMESMANSHIP

It comes as no surprise that Stephen Potter could devote a whole book to the subject of the part that gamesmanship plays in golf. Over seventy years earlier, Horace Hutchinson was already aware of the possibilities. In Hints on Golf *he wrote:*

If your adversary is badly bunkered, there is no rule against your standing over him and counting his strokes aloud, with increasing gusto as their number mounts up; but it will be a wise precaution to arm yourself with a niblick before doing so, so as to meet him on even terms.

and

Of course in every match your ultimate success will depend largely upon the terms on which you have arranged to play before starting. The settling of these conditions is sometimes a nice matter, needing all the wisdom of the serpent in combination with the meekness of the dove. At such times you will perhaps be surprised to hear a person, whom personally you had previously believed to somewhat overrate his game, now speaking of it in terms of the greatest modesty. These preliminaries once arranged, however, you will find that Richard soon becomes himself again—till next match making begins.

Hutchinson was in fact strongly opposed to the use of official handicaps in match play, believing that the negotiation of the number of strokes to be given was an integral part of the match and one which he clearly very much enjoyed. Stephen Potter would, no doubt, have agreed.

The Swing
'Look out,' he said, just as Big Jim Dougan was about to drive. 'There's a fly on the ball. Stand back and start all over again.'

This ploy from an early gamesmanship school story may seem naïve yet it demonstrates well the truism that the first object of the gamesplay should be to break flow, and the second to introduce non-golf thoughts in the swing and the golfgame.

How many people realize that every part of the swing is associated with irrelevant and putting-off thoughts? It is these irrelevant thoughts, always latent, which the gamesman must try to bring to the surface, however buried and fleeting they may be.

Playing for Money
'What shall we play for?'
'You say.'

The man who says 'you say' is one up. It suggests that to play for half-a-crown would be amusing but that his ancestors, members of White's Club to a man, were equally prepared to stake an estate or a mistress on a game of shove-groat or 'Rock-i'-the-Ring'. Opponent is likely to suggest playing for something decidedly larger than is usual for him. May I tentatively suggest for this occasion a new ploy I am provisionally calling 'To-him-that-hath-shall-be-givenmanship' may be tried? Let slip suggestions that there is wealth in your family. Say 'Have you got a car coming for you?' (suggesting chauffeur *milieu*) or 'Father has been asked to lend his Bernardino Taddi for next year's Quattrocento Italian exhibition at the R.A.' (picture worth £100,000). This will bring in the unbreakable rule of money play:

> If stake is more than mother says
> Ah then 'tis you it is who pays.

BASIC

The Drive

Of all the problems which face the golf gamesman, the problem of pure good play is the most difficult to fight. In particular, some of the best gamesmanship brains in America, many of them drained from England which drained them from Scotland, have been bent to the problem of how to be one-up on the man who hits the longer ball.

In normal circumstances it may be possible, for instance, to give advice to a man who is 2 or 3 up: but it is difficult indeed if he is outdriving you. A list of attempted ploys looks little better than a confession of failure. There is the driver from the head of which you unbutton a head cover marked with a large 'No. 4'. There is the remark, if your own drive of 150 yards happens just to have cleared the rough on the right, that 'position is the point here, not distance'.

Then there is the old ploy, first mentioned by me in 1947, of giving your Vast Distances man a caddy who never says 'good shot' but often points to a place, 30 yards ahead, which was reached by Byron Nelson when he played the course in 1946, or, better still, by J. H. Taylor, when he played there with a gutty in '98.

The problem will be solved in time. Funds for our Long Ball Research Wing are welcome and needed. Meanwhile let me give one piece of general advice. Never, never comment on the fact that your opponent has got distance. Never say 'You certainly powdered that one'. Puzzled by your silence, long driver will try to outdistance even himself until, inevitably, he ends up out of bounds.

But the important point to remember is that superiority in length is a myth, or is at any rate cancelled out by relativity. It depends on the standard of measurement. The man who is outdriven at Sandwich can always say 'when Sarazen won here he never used more than a 3 wood'. A following breeze may help you to make the 200 yard mark down wind at the 18th at St Andrew's. But if your opponent beats you by his usual fifteen yards it is usually safe to say:

'Amazing to think that in these conditions Nicklaus *reached the green* in all four rounds of the open.'

Safe unless you are up against a St Andrew's type gamesman who will probably say:

'Yes. I wonder what club he used from the tee. After all, two generations ago Blackwell reached the *steps leading up to the clubhouse* with a gutty.'

It might be added here that the inferior player should never, never in any way behave differently, let alone apologize, because he is inferior. In the days when I was genuinely young and had muscles like whipcord I used to drive nearly 210 yards on the downhill hole at Redhill. My father's best was 140 yards. As soon as he had struck one of these hundred-and-forty-yarders, he would stand stock still gazing after the ball till it had stopped and then pace the distance, counting out loud, and ending in a crescendo 'a hundred and thirty-eight, *thirty-nine*, FORTY'.

It is worth noting here that if Long Handicap is playing Short— 14 playing 4, for instance—never must 14, if he wins, admit, recall, apologize or refer in any way to the fact that he has received 8 strokes in the round; and it is most unusual to refer to this when telling the story to family, particularly wife. This situation and its handling shows yet once again the deep relationship between life and golf, of which life is so often the metaphor or mime.

Style
This is the place to say something not about the style of gamesmanship but the gamesmanship of style. A perfect, flowing, model style can be alarming to an opponent. The teaching of golf is not

our domain: but the teaching of style comes very much into our orbit. An appearance of a strong effortless style, flowing yet built on a stable foundation, can be alarming to an opponent even if it has no effect on one's shots.

'Right, let's have a game then,' says Jeremy Cardew to comparative stranger after a dinner party.

'I haven't played for ages,' he goes on. Though in full evening dress, he may pluck a bamboo stick from a pot in the conservatory and begin to take a practice swing, left hand only.

'My, what a wide arc to that swing,' thinks Wiffley, who is already wondering if he, too, ought to have worn a white tie instead of a black. We recommend the suggestion of great width, on this backswing, and long relaxed follow-through.

Above all we recommend practising a practice swing which ends with the body turned correctly square to the direction of the ball, the hands held high, an expression of easy confidence on the face, a touch of nobility, as if one were looking towards the setting sun. Students who find themselves unable even vaguely to simulate a graceful finish may do well by going to the opposite extreme. It is possible to let go of the club almost completely at the top of the swing, recover it, and by a sort of half-paralysed jerk come down again more or less normally. Opponent will find himself *forced to stare at you*, and may lose his rhythm.

Straight Left Arm: A Personal Confession
I am sometimes asked which, of all the gambits I have invented, do I personally find most useful. Here, exclusively and for the first time, let me reveal the answer to this question.

In golf I have no doubt. Described in Gamesmanship, it is for use against the man who is driving farther and less erratically than yourself.

'I see how you're doing it,' you say, 'straight left arm at the moment of impact, isn't it? Do you mind if I stand just here and watch?'

In spite of the fact that the left arm is always straight at the moment of impact, this used to cause a pull in the old days. Now there is a well-developed counter. But in '68 I am still finding it useful.   STEPHEN POTTER, *The Complete Golf Gamesmanship*, 1968

GREAT PLAYERS

❖

THE GREAT TRIUMVIRATE

J. H. Taylor, Harry Vardon, and James Braid, the great triumvirate as they were called, literally brought golf into the twentieth century. Not only did they dominate the game from the mid 1890s almost until the war, but they raised the standard of play during that time to a level previously unknown. The successes of Taylor and Vardon also marked the first prolonged challenge to the domination of the Scottish professionals and so the true beginnings of the worldwide spread of the game.

J. H. TAYLOR

. . . Taylor's first coming in 1893 did, I think, mark an epoch. He did not win the championship that year. The champion was another young player, Willie Auchterlonie, now the universally respected professional to the Royal and Ancient Golf Club. He would very likely have won more championships had he not devoted himself to becoming a famous club-maker, and no better example could be found than his of the fine, old slashing St Andrews swing. Still, it was this young Englishman, Taylor, from Westward Ho! with his rather curtailed swing, who looked as if he were playing iron shots up to the pin with his brassy, who set all tongues wagging at Prestwick. It is sometimes said that these full shots of Taylor's up to the green were in the nature of a completely new revolution. I doubt if that is quite true. I think Mr John Ball, three years earlier, had first opened men's eyes to the possibilities in this direction, which made vain the old advice of 'Take your cleek for safety.' Taylor, however, rubbed the lesson in, and, moreover, being a professional he played on many courses and so was seen by many more spectators than was Mr Ball, placidly pursuing his own avocations at Hoylake. Taylor's arrival, too, synchronized with a great advance in the general popularity of golf, and so, in one way or another, it is something of a landmark in golfing history.

[69]

It would have been that, I dare say, at any time, because there has been no more remarkable man and golfer. The story of J. H.'s life might be called, like that of the Mayor of Casterbridge, the 'Story of a Man of Character.' For almost innumerable years he has been the acknowledged head of his profession and it is due to him, more than to any other one man, that that profession has climbed so far above its old unsatisfactory condition. He is a natural speaker, a natural fighter, a natural leader, who would have made his mark in any walk of life. With his name are, of course, always associated those of the other two members of the 'triumvirate,' Braid and Vardon; but Taylor, though actually the youngest of the three, was the first to appear; indeed, in his first year or two, Taylor and Herd were the two names bracketed and constantly to be seen together in exhibition matches. . . . To one who had seen only Scottish professionals, Taylor's method seemed strange and new. He put his right forward, and all the books told us to put it back; he stayed almost throughout the swing on the flat of his feet; he did not seem to follow through very much; he hit the ball much as most people hit iron shots and with a formidable little grunt. And of course he hit it hideously straight, so that the joke about the guide-flags being his only hazards became instantly comprehensible.

I am inclined to think that this appallingly and monotonously straight driving of Taylor's has done at least as much as any other part of his game to bring him his five wins in the championships, and his almost numberless seconds. His mashie play has been more lauded, and no doubt it was extraordinarily accurate, though not more so than Vardon's; but I doubt if ever any one over a course of years was so ruthlessly straight in the long shots. He might doubtless have hit further in a freer method, but what little he lost in length he gained in accuracy, and there was no better player in bad weather. When he pulled down his cap, stuck out his chin, and embedded his large boots in the ground, he could hit straight through the wind as if it were not there, and I think the finest golf I ever saw was his at Hoylake in 1913. The rain poured and the wind whistled, and he beat the next man by eight strokes after, incidentally, he had only succeeded in qualifying by holing an eminently nasty putt on the last green.

This was characteristic of him in that he was never so dangerous as after he had fought his way through a bad time. A highly strung man, he suffered cruel tortures, and now and again his emotions might be too much for him, but, when he had mastered them after a struggle, he was indeed terrible, and it is noteworthy that when he won a championship he won it easily. Moreover, however much he suffered, his was an indomitable spirit which rose on the day of battle; he might hate the game but he would never give in to it.   BERNARD DARWIN, *The Golfer's Companion*, 1937

As well as being a fine player, Taylor was obviously a more than useful diplomat.

GOLF PROS

I wonder why all the golf champions that I have known have always been most agreeable men—that is, the professionals. The amateurs, on the other hand, have not always deserved such high praise.

Take James Braid, Sandy Herd, or J. H. Taylor—where will you find better men? I know their companionship does me good, and particularly what I like about them is their keenness—it must be an infernal nuisance for them to play with a bad golfer like myself, yet somehow they give one the impression that they are just as anxious to beat me as if they were playing for the championship.

Once upon a time I went down to Sandwich with the Aga Khan and J. H. Taylor. His Highness was not in form, and was getting depressed. It was summer, and the ground was hard, and as we approached the last green the Aga Khan hit his ball hard on the top and accordingly was stricken with sadness.

Suddenly I heard Taylor's voice saying, 'Really, your Highness, what is the use of my dinning into your ears to pitch the ball up to the hole when these run-up shots come so naturally to you?'

The next day the Aga Khan played very well, and I could not help thinking that Taylor's words of encouragement on the previous day had something to do with it.

                          LORD CASTLEROSSE, *Valentine's Days*, 1934

HARRY VARDON

Vardon stood exactly five feet nine and one quarter inches and during his prime, he weighed around 155 pounds. His hands though were large enough to have belonged to a man twice his size. While gripping a club, his fingers looked like a bunch of sausages wrapped around a baton. With them—for Vardon's power emanated almost totally from his hands—he performed feats of golf seldom seen before or since.

Separating what actually was seen from what only was heard of Vardon has always been a task. But there are some reasonably authentic anecdotes that can be plucked from the endless apocrypha of his life. Here are three.

In a grudge match for £100 against Willie Park 60-odd years ago, Vardon halved ten straight holes against the then-invincible Park and, in the process, dropped his tee shots within a few club-lengths of Park's on every hole. The stalemate was broken on the eleventh fairway when Vardon hit Park's tee shot on the fly. Visibly shaken —as well he might have been—Park never managed to win a hole.

On another occasion, Vardon sank a mashie (about a five-iron) on his second shot to a par-four hole. For those in the gallery who thought the shot might have been due entirely to luck, Vardon managed that afternoon to sink another mashie on the same hole.

While engaged in a lengthy exhibition tour of the United States in 1900, Vardon was hired to display his form by hitting balls into a net at Jordan-Marsh, the Boston department store. To relieve the monotony of what Vardon thought to be an inane way to show off his skill, he amused himself by aiming at the valve-handle of a fire extinguisher that was projecting through the netting. Since the handle was no larger than a silver dollar, hitting it at all would have to be classified as quite a feat. Vardon, however, hit it so often that the floor manager begged him to stop for fear of flooding the store.

Harry Vardon was, to state the situation mildly, a hell of a player. For more than 20 years, or a third of his life, he dominated competitive golf not only in his own country but throughout the world. He is the only man ever to have won six British Opens, and he won them at a time when that tournament was beyond all peradventure the premier championship in the whole of golf.

He won the United States Open in 1900 and tied for first in 1913, only to be beaten in a play-off by 20-year-old Francis Ouimet, the Tom Swift of the fairways. In 1920, at the age of fifty, he came close to winning the title again despite two past seizures of tuberculosis which seemed to have left him without even a vestige of his putting stroke. 'At the time, Vardon was the most atrocious putter I have ever seen,' recalls Gene Sarazen. 'He didn't three-putt, he *four*-putted.'

Harry Vardon was a golfer *au naturel*. He never asked anyone to give him a lesson and nobody ever had the audacity to offer him one. With all his native talent, he nevertheless knew in his mind what every muscle in his body was doing.

Unlike most mechanical golfers, Vardon was an utter stylist, a swinger in the classic sense who employed alterations in form distinctly his own. For example, he advocated an open stance for every shot in the bag 'because,' as he explained, 'there is nothing to impede the clubhead in coming through first.'

'To accuracy,' says Bernard Darwin, the British essayist, 'he added something more of power which put him for a little while in one class with all the other golfers in another. He could reach with two shots greens which asked for two shots and then a pitch from nearly all the rest, and his brassie shot was likely to end as near the pin as did their pitches.'

'Vardon always played well within himself,' says Sarazen. 'He always kept 20 yards of his power in reserve and there were times, when he hit the ball flat out, he could add as much as 50 yards to a drive.'

'He was the epitome of confidence. Often, he would play a shot and then replace his divot before bothering to see where the ball had gone. He didn't have to see. He knew.'

Imperturbable, taciturn, he seldom smiled on the course, but then he seldom scowled, either. He never threw a tantrum, never gave an alibi. He just came to play.

Emotionally, Vardon had the equanimity of a plow-horse. 'If a dog crossed the tee in front of him while at the top of his swing,' wrote Andra Kirkaldy of Vardon in his autobiography, 'he would be able to judge whether the dog ran in any danger of its life. If it did, he would stop his club; if it didn't, he would go through with the shot, without pulling or slicing.'

Harry Vardon was born May ninth, 1870, in Grouville, a whistle-stop on the Channel Isle of Jersey, just off the western coast of France. His father was a gardener. Harry had two sisters and five brothers, two of whom also became professionals. Harry and his brothers took to golf by caddying at a nearby course which had been constructed by some visiting sports from England. He was then eight years old.

For fun, he and his brothers and some other caddies built a course of their own, consisting largely of 50-yard holes. They fashioned clubs by using oak branches for heads, with strips of tin for faces, and black thorn branches for shafts. Marbles were used in lieu of golf balls.

As his father had before him, Harry became a gardener in his teens. His first job was with a retired army major who was a fanatic on golf. He encouraged Vardon to play and gave him his first set of clubs.

Through one of his brothers who had become a professional, Harry secured a pro berth at the Studeley Royal Golf Club in Ripon, Yorkshire, while in his early twenties. In 1896, at the age of twenty-six, he won his first British Open by defeating J. H. Taylor by four strokes in a 36-hole play-off at Muirfield. He won the championship again two years later at Prestwick, repeated in 1899 at Sandwich, his favourite course, and returned to Prestwick in 1903 to win again, this time with the new rubber-cored ball. Part of Vardon's greatness lay in his ability to adjust his game from the gutta-percha ball to the rubber-cored, a change that not more than a handful of other golfers were capable of making.

After the first of two major battles with tuberculosis, Vardon managed to win his fifth British Open at Sandwich again in 1911. His record sixth win was accomplished in 1914 at Prestwick.

The really remarkable fact about Vardon's golf is that he played with only seven clubs: a driver, a brassie, a cleek, a driving mashie, an 'iron', a mashie and a putter, as they were called in the parlance of the day. These were equivalent, in today's terms, to a driver, a brassie, a shallow-faced one-iron, a three-iron, a four-iron, and a five-iron, which he used for recovering from sand traps by laying open the face. Later in his career Vardon added three clubs, one of which was a left-handed mashie, or five-iron, which he used when fences or trees interfered with his normal stance. There are those who say there was no perceptible difference between the results of his left-handed mashie shots and his right-handed ones. He was capable of tearing the flag out of the hole with either.

CHARLES PRICE, Golf, 1960

❖

JAMES BRAID

Braid's is not altogether an attractive style to watch. Sound it certainly must be, or it could not execute its results; but one would not call it orthodox. A point to be noticed . . . is the comparative shortness of the swing, computed by the position of the club when at its highest. It is apparent that it does not nearly reach the horizontal behind the back—scarcely is it allowed to make more than an angle of 45° behind the back. We reckon this as not enough for orthodoxy; perhaps we may regard the horizontal as about the proper standard of the classic style . . .

If he has rather less than most behind the ball, it is certain that he has a deal more than most before the ball. After the ball is hit he seems to be putting his greatest force into hitting it; and this is probably the main secret of his length of drive. The attitude in which he is shown after the stroke, facing right towards the line of flight of the ball, shows how freely he has let his body follow on with the stroke. His right leg seems left right away behind him; in his upward swing, even with his short back stroke, the weight of the body had come on the right leg, but now the body is right away forward on the left leg. The right shoulder has come round fairly far, but the most noticeable point about the swing is the manner in which the body with the arms and the club have been let come right away through—almost too much so, to say the truth, for grace. Braid's stroke is not altogether a graceful one—he will forgive us we hope, for the criticism—but its terrific power is absolutely undeniable. No player that we are acquainted with is able to indulge himself so safely in the generally dangerous luxury of pressing. He seems to press hard with every drive he makes; but the true meaning of this is that he is in such perfect training of eye and muscle that he is able to concentrate all his force on the stroke in a degree that it would be impossible for any man in less perfect training of eye and muscle to combine with a sufficient degree of accuracy. The great feature of his style is his power of keeping such tremendous force under such perfect control.

Therefore, in spite of its lack of grace, and partly by reason of its lack of grace, it is a peculiarly fascinating style to watch. This sounds a paradox; but there is a special delight in seeing the kind of divine fury with which he 'laces into' the ball, and yet the wonderful accuracy with which the club meets the ball. He is a fine player to watch. HORACE HUTCHINSON, *Golf and Golfers*, 1889

❖

VARDON AND RAY AT BROOKLINE

I suppose I never had the least notion that golf offered more than merely personal competition, as a game. Even when we had a medal round, I was trying to beat Perry or whoever seemed the most dangerous competitor. I kept my scores in every round of golf as a matter of course; and naturally I liked to get under 90. But the scoring, of itself, was relatively a detached part of the affair, which, to my way of thinking, was a contest with somebody—not with something.

It may not be out of place here to say that I never won a major championship until I learned to play golf against something, and not somebody. And that something was par. . . . It took me many years to learn that, and a deal of heartache.

It was in 1913 that Harry Vardon and Ted Ray came over from Britain to play in our national open championship at Brookline, and Francis Ouimet, then a boy of 19, shot his way into a triple-tie with the famous English professionals and beat them in the playoff. That is the first golf I remember reading about in the papers, and I began to feel that this was a real game. In October, Vardon and Ray played an exhibition match at East Lake with Stewart Maiden and the late Willie Mann, then professional at the new Druid Hills Golf Club in Atlanta, and defeated our boys 1 up, Ray having to sink an eight-foot putt on the thirty-sixth green to get a half and win the match.

This was the first big match I had ever watched and I followed every step of the 36 holes. Ray's tremendous driving impressed me more than Vardon's beautiful, smooth style, though I couldn't get away from the fact that Harry was scoring more consistently. Par—par—par, and then another par, Harry's card was progressing. They played 36 holes at East Lake and 36 more at Brookhaven next day, and I remember Vardon's scores: 72–72–73–71; a total of 288, or an exact average of 4's all the way. I remember thinking at this time, and it would be difficult to find a better illustration, that 4's seemed to be good enough to win almost anything. Today I would not qualify the estimate. I'll take 4's anywhere, at any time.

On the twelfth hole of the afternoon round at East Lake, Ted Ray made a shot which stands out in my mind today as the greatest I have seen.

Our boys had finished the morning round 2 down but had started brilliantly after luncheon, especially Stewart, and had got in the lead. Then, beginning with the twelfth hole, the visitors executed four birdies in succession and went back in front. Vardon got the birdie at No. 12, but Ray, in getting his par 4, produced this astonishing shot. His drive was the longest of the four, as usual, but right behind a tree. The tree was about forty feet in height, with thick foliage, and the ball was no more than the tree's altitude back of it, the tree exactly in line with the green. As Ray walked up to his ball, the more sophisticated members of the gallery were speculating as to whether he would essay to slice his shot around the obstacle to the green, 170 yards away, or 'pull' around in on the other side. As for me, I didn't see anything he could do, possibly; but accept the penalty of a stroke into the fairway. He was out of luck, I was sure.

Big Ted took one look at the ball and another at the green, a fair iron-shot away, with the tree between. Then without hesitation he drew a mashie-niblick, and he hit that ball harder, I believe, than I ever have seen a ball hit since, knocking it down as if he would drive it through to China. Up flew a divot the size of Ted's ample foot. Up also came the ball, buzzing like a partridge from the prodigious spin imparted by that tremendous wallop—almost straight up it got, cleared that tree by several yards, and sailed on at the height of an office building, to drop on the green not far from the hole. . . . The gallery was in paroxysms. I remember how men pounded each other on the back, and crowed and cackled and shouted and clapped their hands. As for me, I didn't really believe it. A sort of wonder persists in my memory to this day. It was the greatest shot I ever saw.

Yet, when it was all over, there was old Harry, shooting par all that day and the next. And I couldn't forget that, either. Harry seemed to be playing something beside Stewart and Willie; something I couldn't see, which kept him serious and sort of far away from the gallery and his opponents and even from his big partner; he seemed to be playing against something or someone not in the match at all. . . . I couldn't understand it; but it seemed that way.

R. T. JONES, *Down the Fairway*, 1927

❖

BEN HOGAN

In any discussion as to the greatest golfer of the past half-century the names of Bob Jones, Ben Hogan and Jack Nicklaus stand alone.

[77]

To attempt comparison between them, absorbing though it may be, is a purely academic exercise because far too many variables are involved. No man can do more than achieve indisputable supremacy in his own time, and thereby acquire an aura reserved only for those who do something better than anyone else in the world. However familiar as persons these supreme beings may become they quicken feelings far surpassing normal admiration for great skill, feelings of awe, even wonder, that they are not as other men. Never was this feeling more intense than when one was with Hogan.

There was a rare sense of the unforgettable about him, not simply because of his matchless skill but because of the force of his personality, a force more compelling in its fashion than the charm of Arnold Palmer, the ebullience of Lee Trevino or the amiable strength of Jack Nicklaus. There was about Hogan in his competitive years, as of no one else, an almost overwhelming sense of a man apart.

Within seconds of meeting him for the first time anyone would be aware that he was in the presence of a most unusual man, even without knowing of his eminence as a golfer. The outlines of his face, indelibly tanned by thousands of days in the sun, are unmistakably powerful, even cruel when the wide mouth is drawn down in concentration; the piercingly direct gaze could be cold as a winter dawn, and his economy of speech fearsome. In the tradition of the Far West Hogan does not believe in wasting words.

Even in a longish conversation I have never heard Hogan waffle; his replies to questions were invariably direct, sometimes alarmingly so; where a monosyllable would suffice he saw no reason to cloak it in useless verbiage. I recall waiting outside an airport for transport to a course some years ago; Hogan and his wife, Valerie, were there and eventually a bustling official approached and asked Hogan if he would care to ride down with Mr Snead. Hogan looked straight at him and said quietly, 'Not particularly.' Not another syllable did he utter and the little man scarcely knew where to look. The Hogans rode alone.

The American Press as a whole are not famous for their deep insight into golf or for watching it and the players can be asked stupid questions. On one such occasion Hogan is said to have remarked that, 'One day a deaf mute will win a tournament and no one will know what happened.' Another time I was talking with him when a man approached and said, 'I suppose you come to Augusta now to see your friends,'—a lethal misunderstanding. Hogan chilled him with a glance and replied, 'I see my friends in Fort Worth.' On the other hand, if Hogan knew that a writer was genuinely interested, and watched the golf, he would take pains to explain the shot.

Tournament golf was not a social occasion for Hogan. He would enjoy chatting and having a drink with old cronies in the locker room, but even in the later mellower years one would never see him pausing here and there about the clubhouse or its precincts. It was as if he had need of no one, either on the course or off it, except for his wife; he seemed a man alone and I imagine sometimes a lonely man.

This very self-sufficiency indicated the force of character that had enabled him to endure and overcome privation and adversity, the like of which none of his great contemporaries have known. As a child he sold newspapers until he heard that he could make 65 cents a round as a caddie. He had to fight to get a place and said that he would run seven miles to the course. Eventually he had one club, a left-handed mashie, and used to tear up the grass in the back yard until his mother would send him to the grocery store and he would hit shots along the road. When he changed to right-handed he said he had the most awful time, and that the effort used to nauseate him.

Hogan started professional golf when he was 18, and all the world knows of the years of struggle and hardship before he could win. His first major victory, the PGA Championship, was not until 1946 when he was two years older than Nicklaus is now. Then, on the threshold of absolute supremacy, came the accident which nearly destroyed him, leaving its mark forever afterwards and compelling him to husband his resources and not tire himself socialising. The ensuing remoteness helped to compound a figure of legend, as indeed he became to a greater extent than any other golfer of his time. You could feel it whenever he appeared.

If Hogan walks into a room you are aware of a presence, although nothing of his manner or outward appearance attracts attention. Always he dresses impeccably, inconspicuously, in modest shades. Vivid colours and cheap gimmicks are not for Hogan; there is nothing about him of the extrovert effusiveness so common in the American male; always in public his manner is quiet and contained.

More often than not in those later years he would be addressed as Mr Hogan; others were Arnie, Billy or Sam on the briefest acquaintance, but Hogan was Mr and the title is more meaningful in America than in Britain. Similarly, on the course Hogan never sought acclaim for its own sake, or in any way played to the gallery. An army of screaming fans, like those on which Palmer thrives, meant nothing to him. He allowed his masterful golf to speak for itself and it usually did. The applause that greeted him as he approached every green had a note of respect reserved for no one else; the people, reverent rather than ecstatic, would clap rather than cheer.

There is no doubt that Hogan came closer than anyone to eliminating the human element from golf. Such was his control of the ball, and completeness of technique, and so true in outline and rhythm was his swing that he seemed immune to the pressures that destroy other golfers. When he was ahead in his great years his command of an event seemed unbreakable; rarely if ever was there any stumbling on the way to victory, such as that which has troubled Nicklaus on occasion and Bob Jones long ago. One of the few regrets that Jones had about his competitive golf was that he sometimes surrendered a commanding position and won far less easily than he should have done. Not so Hogan: as Jones himself said, 'Ben was always so good at finishing the job.'

Probably the greatest instance of this was in 1951 when Oakland Hills had been made savagely penal for the United States Open. Clayton Heafner with a 69 was the only other player to break 70 in the whole Championship. Hogan's scores read 76–73–71–67. The final round conceivably is the greatest he ever played, although when I asked him to confirm this years later he said: 'It is very difficult to get your mind back on to one particular round, analyse it again and say whether it was better than any other one.' This was revealing of the man in that he would estimate greatness by the quality of the shots rather than the winning of a championship. He went on to mention a round he had recently played, some of which I saw. Laurel Valley that day of the PGA Championship was very long and heavy, but he missed no fairways and only one green by three yards, and was round in 70 having taken three putts five times. He thought he had played as well as ever he could; he was then 53.

No modern golfer has ever finished the job more emphatically than Hogan did in 1953. He won the Masters by five strokes with a total of 274 that only Nicklaus has beaten; the United States Open by six and the British by four. His golf at Carnoustie was so masterful that had his putting been fractionally better than moderate the Championship would have been no contest at all. His scores were 73–71–70–68, another ruthless downward progression, and one felt that had a fifth round been necessary it would have been 66. This was the peak of his supremacy.

Hogan won no more championships, but 10 years later the professionals were still saying that from tee to green he could hit the ball better than any man alive. But even Hogan, tough of spirit as he was, did not have an inexhaustible supply of nerve and gradually putting became an agony and a frustration to him and the worshipping watchers. Leaving the practice ground one morning at Augusta on his way to the putting green he said, 'Now for the bloodbank,' and he meant it.

In the third round of his last appearance in the Masters in 1967 Hogan, rising 55, played the back nine in 30, and the only putts holed of any length were from six, 15 and 25 feet. For thousands a lifetime's memory had been made, but the next day brought tiredness and reaction.

Hogan was the ultimate perfectionist, the more remarkable because he emerged from an age when intensive practice was not the fashion. Once, after a 64 in a tournament he practised for two hours and everyone thought 'I was nuts.' He practised with a ferocity of intent that can never have been surpassed, as if refusing to believe that it was not possible to create a flawless swing. There can never have been a more acute golfing intelligence than Hogan's, and for a great part of his lifetime he concentrated it upon the search for perfection.

I have the impression of him that hitting pure golf shots was the fulfilment of his whole being, as much an expression of his soul as a painting to an artist, a piece of music to a composer, and he was never satisfied. He saw no reason why every hole should not produce a birdie, and I could imagine his feelings when he dreamt one night that he had 17 holes in one, and one in two. Telling this he turned to me almost angrily and said, 'When I woke up I was so goddam mad.'

The pursuit of perfection was not confined to the golf course. When he began to manufacture clubs in 1954 he was so displeased with the output that he insisted on having it all destroyed. Nothing but the finest would appear under his name. The gesture cost the company a great deal of money, but Hogan's pride was satisfied.

The swing he created was not beautiful except in the sense that there is beauty in the smooth working of any machine. The backswing was compact and flatter than the classical concept, but the extension of the arc through the ball was quite remarkable with the right arm remaining straight through to the finish of the swing. The tempo was fastish; it was remote from the lazy effortless grace of Jones or Snead, but there was about it a wonderful sense of precision, like an instrument of flawlessly tempered steel.

For many years his accuracy was phenomenal. He could shape the shots as he willed, and even late in his career he could place drives as precisely as others would medium irons. To watch him fade or draw the ball ever so slightly, usually away from a hazard, was an unforgettable sight. The flight of his long irons had a searing quality, the ball so truly struck that it seemed motionless as if drawn down a plumb-line. If accurate striking be the measure of greatness then Hogan must be classed as the finest of all golfers. Gene Sarazen, who has played with every great player of the half-century since Vardon has often said that no one ever covered the flag like Hogan.

In 14 consecutive United States Opens up to 1960, and the same number of Masters to 1956 he was never outside the first 10, and was in the first four 18 times.

Of all the impressions of Hogan that memory cherishes those of him against the matchless background of Augusta are perhaps the strongest; many of the tees are cool places of green gold shadows, cloistered in the trees. Hogan, swarthy, silent and inscrutable, appears, tees his ball and takes his stance with that impression of massive authority created only by the great ones; the powerful hands, mahogany dark, mould themselves on the club, the swing coils and uncoils, swift and true as a lash of steel; the ball arrows away and Hogan with his stiff rather limping walk moves out of the shade. A moment later he stands in the sunfilled loneliness of the fairway. At such moments there was a rare quality of stillness about him and one could sense the absolute concentration of the cold, shrewd mind as he surveyed the shot.

PAT WARD-THOMAS, *Piccadilly World of Golf*

WALTER HAGEN

There must be more stories about Walter Hagen than about any other golfer before or since. He was clearly the master of match play and a ruthless competitor, whose ability to recover from apparently hopeless positions must have been as dispiriting for his opponents as his often cruel gamesmanship.

Bernard Darwin describes him in an altogether more friendly way than some of his opponents would have done. Perhaps the truth lies somewhere between this view and that suggested by the accounts of his match with Sarazen and Byron Nelson's first encounter with him, both of which are to be found in the section on tournaments and matches.

In 1920 there first appeared in this country at the age of 28, Walter Hagen, not only a very fine player but an extraordinarily picturesque figure, a godsend to those who like to write 'colourful' descriptions and retell anecdotes. He was a man as to whom it was hardly possible to take a neutral view; you might like or dislike him but he compelled your attention; his character and flavour were such that they could not be neglected. The sum of his golfing achievements is so large that it is only possible to set down a few of them, and in any case the man himself was to most people more interesting than his feats; he had such a way with him that crowds were ready to watch him when he had not the remotest chance of winning. In fact apart from numberless big tournaments in his native country—and he had the happy knack of winning when there was the largest number of dollars at stake—he won two American championships and four British ones. Yet I am inclined to cite as his most remarkable feat the winning for four consecutive years of the American P.G.A. tournament. That corresponds to our 'News of the World' tournament and is played for by matches and, considering the strength of the opposition, to win it four years running (he won it another year as well) was truly astonishing. It showed more than anything else that power of dominating, almost of cowing his rivals which was one of his strongest assets.

There have been, I think, more skilful and certainly more mechanical and faultless players than Hagen, but none with greater sticking power or a temperament more ideally suited to the game. He was a strange mixture of two usually contrasted elements, on the one hand the casual and the happy-go-lucky, on the other the shrewd, long-headed, observant and intensely determined. His manner while playing was a reflection of his nature, for he could 'let up' between strokes and converse in a carefree manner with a spectator and then switch off this mood and switch on one of single-minded attention to the next stroke to be played. No doubt while he was talking with such apparent *insouciance* his mind was busy looking ahead, but he had an almost unique power of relaxing and never presented that aspect of stern and solemn pugnacity without which the less happily gifted cannot concentrate their minds.

Much, too much, has been written of Hagen's gifts as a 'showman' and no doubt he fully understood that his casual manner, with a touch of flamboyant swagger about it, went down well with the crowd. No doubt also he could and did turn it on as some people can turn on, sometimes too palpably, their charm. But the casualness was natural to him and not forced. It helped him to take a strenuous life easily and unexhaustingly. It likewise enabled him to run things fine in point of time and even to be late in a way which might exasperate other people but did not cause Hagen himself to turn a hair. Others might wish to lie in bed or to stay in their baths, but they could not do so because of some malignant sprite whispering in their ears that they might have to hurry to the tee or even be disqualified. Hagen stayed in his bath as long as he pleased and trusted to the chapter of accidents. Possibly he was a rather spoilt child in this matter in that promoters of tournaments knew his value and were always ready to make allowances for what was 'pretty Fanny's way'. There were occasions, notably in a certain match here against Abe Mitchell, when Hagen did considerably ruffle everyone by this inconsiderate lateness. There were not wanting those who said that it was part of a deep-laid scheme to disturb his adversary, much as in older and less scrupulous days men would deliberately fret a nervous adversary by breaking away at the start of a hundred yards' race. Personally I think such accusations were utterly unfounded. I do not for a moment believe that he had any such design; this would not have accorded with his code; he was just irretrievably casual and had the bump of punctuality, if there be such a thing, very imperfectly developed.

That Hagen had an overpowering effect on some of his opponents was clear enough. His demeanour towards them, though entirely correct, had yet a certain suppressed truculence; he exhibited so supreme a confidence that they could not get it out of their minds and could not live against it. They felt him to be a killer and could not resist being killed. He had a very shrewd eye for their weaknesses and, strictly within the limits of what was honest and permissible, he would now and then exploit them to his own advantage. I heard a story the other day of Hagen's tactics which seems to me eminently characteristic, and I believe it to be true. He was playing in one of those four consecutive finals of the American P.G.A. tournaments which he won, and with one hole to play the match was all square. Hagen having the honour, sliced his drive and the ball sailed away into a wood on the right while his adversary went rigidly down the middle. Hagen carefully examined his ball and emerged from the wood for a minute to have the crowd moved back, as if he were going to make the best of a bad job and play out sideways. He went back into the wood, had another look and then, as if suddenly spying a loop-hole of escape, played a magnificent iron shot through a gap in the trees right on to the green. The flabbergasted enemy put his ball tamely into a bunker and the match was over.

BERNARD DARWIN, *Golf between Two Wars*, 1944

❖

HAGEN AT ST GEORGE'S

I picked up Walter Hagen at the Canal. He had the biggest crowd, including the Prince of Wales whom I wormed my way next to in the front row round the green. Hagen was wearing black-and-white shoes, dark blue stockings, dark green plus-fours made of alpaca, a dark blue sleeveless pullover, a white shirt with gold cufflinks and a dark blue-and-white bow-tie. His black hair was patent leather. The feature of his brown face was a sharp long probing nose. Most of our professionals played in drab shabby old things. The conception of dressing up for golf was new; there had been gibes at Hagen in the press. Now that I saw him I was immediately on his side. I watched his drive and trotted as near as I dared to him. He strolled along laughing, joking, and laying bets. About twenty yards from his ball he fell visibly silent, his gossips took their cue and he advanced, now, rather than walked. His head was erect and still, his nose like a pointer's. He seemed to me to send his eyes physically forward to the distant green, and they reconnoitred minutely the slopes, the position of the guarding bunker, indeed a whole limited area, and I knew what he was doing: he was deciding whether to play safely short, or 'go for the green'. This meant pitching a full brassie shot within a minute area. All this time he was moving on

> With unperturbed pace,
> Deliberate speed, majestic instancy.

It was an extremely dramatic moment. As he reached his ball the crowd congealed into that almost tangible cold silence unique to the golf links. I was directly behind the line. Hagen selected his brassie with a quick certain flourish that got a gasp of awe, stepped straight to his ball and in a second had hit it. I watched it the length of its flight—the straight rocketing upward trajectory, the beginning of the earthward curve, its steepening fall, the pitch and bounce up. The frozen moment melted. There was a cheer from the green, a scurrying surge of legs, and chatter, and Hagen strolling on, laughing, joking and laying bets. He won the championship. I had followed his play before, as far as I was able to; now I idolised him. He was with Wethered and Tolley.

PATRIC DICKINSON, The Good Minute, 1965

BYRON NELSON AND JOHNNY MILLER

Every now and then a golfer will achieve a level of performance which for a time seems to make him virtually unbeatable. Probably the two greatest winning streaks were those enjoyed by Byron Nelson in 1945 and Johnny Miller in 1975. The following extracts tell their stories in their own words.

BYRON NELSON

The Fort Worth *Star-Telegram* published at the beginning of the summer season all the tournaments over the whole state, and there was one you could play in absolutely every week—amateur tournaments, that is. That's one reason why Texans kind of dominated the game for a time. And another thing about Texas was the wind blew a lot, so you had to learn how to cope with it. The wind still blows a lot, but in those days you also had hard greens, and had to play off hard fairways. You really had to learn how to manipulate the ball. When we got off Bermuda grass and went to bent, boy, that was heaven.

We don't produce golfers down here like we used to, because everyone in the country plays under the same conditions—watered fairways. So you can say that great golfers are created under bad conditions. Personally, I don't think there's any question about it. It was tough, playing back then. You had to learn to play diddy-bump shots, chips and runs, and on hard greens.

Well, it all came to a head, I guess, in 1945, with that streak. I won eleven straight tournaments, eighteen for the year. But I think it started in 1944. Of course, I'd had some good years before. In 1937 I won my first Masters and the International Match-Play championship. And '39 was a great year. I won the Western Open, the North and South, and the National Open. Anyway, I won six tournaments in 1944, and averaged 69.67. I played very well, and the thing I did, I kept track of every round, made little notations, just like when you take inventory at the end of the year, and I found two things. I chipped poorly too many times, and played too many careless shots. By that I mean I didn't concentrate hard enough. For instance, there'd be an out of bounds on the right and no trouble on the left and I'd just walk up and hit it and get into trouble. Because the truth is when you're playing well you quit concentrating. You can get a little complacent if you just hit it good, good, good. It can get kind of boring when you're playing as well as I was, and then all of a sudden you quit paying attention. I had people say to me it was no fun watching me play, because it was just on the fairway and on the green, two putts, and once in a while I'd make a birdie putt. The average person may not understand that it can get boring for the player, it sounds like a silly thing to say, but it's true. Anyway, I made up my mind to do some practice on my chipping for the 1945 tour, and I said, 'I'm not going to play one careless shot.' Those were the only things I had to change.

Of course, the pressure built up something terrible. I know one time I said to Louise—this was along about the seventh or eighth tournament—that I wished I'd blow up today, I don't care what I shoot, I just want to get rid of this. And I came home and she said, 'Well, did you blow it?' and I said, 'Yeah, I shot sixty-six.' I think I had twenty-five 66s that year.

The thing I'm more proud of than anything else about my playing career is the degree of consistency with which I played. I won money in 113 consecutive tournaments, and you have to realize we had a lot of tournaments where they only had fifteen or twenty prizes, not forty or fifty the way they do now. It showed that my ability to concentrate and keep on playing if I bogeyed a hole was good, and I could still come in with a respectable round of golf. Because it's easy, especially after you've been playing well, to have a bad day and say, Oh well, there's always next week. But, of course, I didn't feel that way.

Is there a psychology for winning? I don't understand the psychological function of the human mind sufficiently to answer that very well, except to say that winners are different. They're a different breed of cat. I think the reason is, they have an inner drive and are willing to give of themselves whatever it takes to win. It's a discipline that a lot of people are not willing to impose on themselves. It takes a lot of energy, a different way of thinking. It makes a different demand on you to win tournaments than to just go out and win money. I know when I first started playing out there, I thought, Well, if I can just finish in the first ten, I can make a living. Because the purses were $5,000 in those days and tenth place was worth $180, and that would get you by for a week if you lived conservatively. But I soon realized that if I was going to try to win only $180, I was never going to win $500 or $5,000.

It's hard to explain about winners, or champions. There's a certain aggressiveness. That's the thing I saw in Tom Watson the first time I watched him play. There is something about the mannerisms. I know the first time I won the Masters, in 1937, I just felt like I had springs in me. I didn't hardly feel like I was walking.

BYRON NELSON, in AL BARKOW, *Gettin' to the Dance Floor*, 1986

JOHNNY MILLER

I traveled to Las Vegas in the winter of 1983 to seek out Johnny Miller, the golfer who in 1974 and 1975 touched the highest peaks of the game. He won eight tournaments in 1974, then the 1975 Tucson and Phoenix Opens back-to-back, shooting twenty-four under par and twenty-five under par respectively. His play was uncanny. Time after time he hit shots the exact distance to the flagstick. His play then tapered off, and he rarely played that kind of golf again. The zone he inhabited in the mid-1970s is a foreign country that he later visited for a few shots at a time, perhaps a round here and there. He wanted the experience again, but it's elusive. Tom Watson puts it this way: 'You can get close enough to mastering the game to feel it, to breathe it maybe, to smell it.

But you can't master it, not for a long time. Ben Hogan may be the only person to have really come close to it for a long time. And also Byron Nelson. He came awfully close to hitting the ball the way it should be struck, and also with the scoring part of the game.' Hogan is still acknowledged as the one man who could control the flight of the ball shot after shot. Nelson won eleven tournaments in a row in 1945; his swing was powerful and elegant and always on track.

Miller was the same those few years ago. Now, seated in a plush, purple lounge chair in a hotel room high above Las Vegas, Miller recalls his play. To hear him speak, one might think he's talking about a person other than himself.

'I had it all then,' Miller tells me. 'I hit the ball well with the long clubs, was great with my irons, and putted excellently. I was into something I guess few players have known, maybe none in my era. It was sort of the golfing nirvana, and it had nothing to do with the mental side of the game. It was totally physical. I'd go out on the practice tee in early 1975, hit twenty-five or thirty balls and say, "Perfect, why am I here?" I'd say my average iron shot for three months in 1975 was within five feet of my line. I was hitting an average of sixteen greens a round. Guys win hitting thirteen greens a round.'

What might account for such precision? Miller answered quickly.

'I had such great eyesight and depth perception. I knew what 147 yards was. I knew exactly the difference between 147 yards and 149 yards, and I could feel it. A lot of people hit the ball on line, but don't get the right distance. I did it by taking every club back the same distance, to the same position in my backswing. It didn't matter if I was hitting a half shot or a full shot. The club went back to the same place, and then my whole thing was varying clubhead speed. That's all I was doing. I would take a five-iron back and say, "Okay, this is going to be 100 miles per hour, or 95, or 90." Most guys try to vary distance by choking down on the club, gripping it lighter, swinging softer or harder, or taking a bigger or smaller shoulder turn. But these things had nothing to do with my swing. It was like a pitcher who would make the same windup and then change only the speed of his delivery. That's what I was doing. I was taking the identical windup with each shot, and even the same follow-through, but just changing the speed at the bottom of my swing.'

Miller's play affected golfers in his group. They were befuddled. But what could they do? He didn't know how he came upon this so-called nirvana, nor could he have advised them where they might find it. Yet he didn't seem to be doing anything much different from what they were. It was just that his swings were more refined, that the ball off the clubface sounded sweeter, that his shots finished closer to the hole. The atmosphere around him was quiet, as if reflecting the simplicity in his mind. To be near him was to feel serene, privy to some magical experience that defined what one could achieve in golf.

'I had more guys blowing shots over the greens,' Miller continued. 'It got quite bad in my prime. I was taking this big long swing and hitting the ball nowhere, just concerning myself with accuracy. I'd hit a five-iron because that was the shot I wanted to hit, a real soft shot. Or I'd use a six-iron into a green that was soft and tilted back to front, take a big swing—although it was way too much club—and hit the ball with no spin whatsoever. It would just knuckle up there, hit the green and go pfft, just staying there. The guys hitting after me would see I'd hit a six-iron with this big swing and say no way, it couldn't be more than a seven. They'd hit the seven and blow it right over the green, or they'd hit a shot with so much spin it would suck back twenty-five feet. I had to tell the guys not to look in my bag to see what I'd hit. After I hit the shot I'd tell my caddie I hit it real soft, just to help them, because I knew they were listening.'

Miller carried himself back easily to his best days, speaking without any hesitation. He was talking, I thought, as he once golfed: with complete freedom and direct contact established between his thoughts, feelings, expressions, and his words. How odd that he could not find this mood with a golf club in his hands now, when it was with a golf club in his hands that the mood was previously generated.

Miller controlled his swing in those days. This was not a control born of willfulness, but entirely the opposite. The golfing muse had descended upon him; he was the favored recipient and talented enough to let it find full expression. He felt connected, via the shaft of the club, to the clubhead, and through the contact he could conjure so readily. From the clubhead to the ball, his body would unwind on the throughswing, as he felt completely in touch with his target. In some way, the club swung him. Miller went along for the ride. It was as if he could stretch himself out all along the fairway and touch the pin. Every golf swing felt slow to him; he felt that he had so much time to complete his swing. A new game—the game we all seek—burst through his normal patterns, and he played by sensation, and sensation alone. No wonder he talks about it now with a sense of amazement.

'It was that I had the means of controlling distance,' he tells me. 'I could feel the shot so well. That gave me more shot options, more arrows in my quiver. Most golfers, say, hit a five-iron 175 yards with their normal swing. If they go right to left with the draw, they'll hit it 180; if they play left to right with the fade, they'll hit it 170. That gives them a ten-yard variance with full swings. That's a good way to vary distance. But the trouble in golf so often is that if the pin is left and you're between clubs, you really don't want to hit the ball left to right, so now what do you do? The draw shot will go too far. I could do more things because I was down to a two-yard variance with clubs, since I could vary club-head speed. It was as if I could hit more notes on the scale. And I think that's the ultimate way of controlling distance, to have a set way of taking the club back, then applying the different speeds. It's like shooting a basketball. You set it the same way for a short shot as for a long shot, but you just don't put as much effort into it.'

Miller's memories continue to tumble forth. He stands up to emphasize a point, then swings his arms round as if he were making a swing. Years have passed since he and the game were so synchronized with one another, but something is coming alive in him now, a deep-seated memory of golfing perfection.

LORNE RUBINSTEIN, *Links*, 1991

❖

HENRY COTTON

When anyone asks me who is the greatest striker of a golf ball I ever saw, my answer is immediate. It is Henry Cotton. I am just old enough to have seen Harry Vardon play, but was not old enough at the time to make a fair assessment of his powers. Whatever they were, I cannot believe them to have been greater than Cotton's in the 'thirties. He lifted up the nation's golfing spirit after eleven long years of American domination and, with it, the status of his own profession.

For this the Americans themselves were largely responsible. In

1928, when he was twenty-one, he set sail for the United States under his own steam, buying his own ticket and taking with him a letter of credit for £300, which incidentally he brought home intact. He soon appreciated that the great sporting figures of the day were regarded in America almost as the aristocracy, whereas at home sport carried with it no special standing. When Walter Hagen came to England to win our championships, he stayed at the Savoy and drove up to the course in a hired Rolls-Royce. He was already 'one up' on the rest of the field. Cotton decided that what Hagen could do he could do.

I think it is fair to say that Cotton regarded himself, in his competitive days, as a kind of 'property', to be taken the greatest care of and kept in the best possible condition if it were to give the desired results. For this reason he took it to the best hotels and at lunch time, having no desire for the smoky air and, for the celebrated, the inevitable attachment of bores and sycophants to be found in the club-house, he changed his clothes in the car and retired to the hotel. Naturally enough, there were those who thought he regarded himself as too good for the common herd.

He developed an immense strength in his hands, and they became the focal point in his essentially simple swing. As the ball flew straight at the flag, you felt that, if you hit it in that fashion, it could hardly do anything else. He could do almost anything with a golf ball on purpose and would have made a great trick-shot artist. We often used to challenge him to take his driver from a bad lie on the fairway, simply for the aesthetic pleasure of seeing the ball fly away as though fired from a rifle, and I remember once at Bad Ems seeing him knock a shooting stick out of the ground with a 1-iron shot at a range of 20 yards. We christened him the Maestro, and he deserved it.

At the same time he developed a flair for getting himself into the news, sometimes, but not always, on purpose. With all this he was naturally the centre of attraction wherever he played, and became probably the first professional golfer to be recognizable at once to the man in the street.

In 1929, now aged twenty-two, Cotton played in his first Ryder Cup match at Leeds, where he beat Al Watrous by 4 and 2, and it now seemed only a matter of time before he won the championship. He had his chances, but on at least two occasions let them slip, mainly, as he now thinks, through listening to the rumours that used to fly about the course before the present walkie-talkie system came into use, and not appreciating what he needed to do.

It was against this background that the championship opened at Sandwich in 1934. Cotton had with him four sets of clubs—why, I do not know—and for once could not hit his hat with any of them. He practised on Saturday till darkness drove him in, and had never been in such discouraging form. He settled on a set of clubs, for better or for worse, and on Monday morning, in the first qualifying round, was drawn to go out first, accompanied by a marker. He played what remains in his own opinion the best round of his life. He hit every green in the right number—33 shots, 33 putts; total 66. Such is golf.

The magic lasted. He opened the championship proper with a 67, and in such a way that one saw no reason why he should ever again take more. On the second day he arrived on the 17th tee needing only two par fours for another 67 and the then fantastic total of 134. At each hole he hit a tremendous drive. His second to the seventeenth ruled the flag and finished about 12 feet from the hole, and he holed the putt for a 3. He hit another magnificent iron shot to the last hole, though he cannot have seen it finish, for he was at once enveloped by a stampeding multitude determined to see history being made on the last green.

I remember the shot perfectly. It bounced a couple of times and came quietly to rest about four feet from the stick. He made no mistake with the putt and history had indeed been made. Sixty-five! A total of 132 and the nearest man, Alfred Padgham, nine strokes behind.

On the morning of the final day Cotton turned in a 72 in harder conditions, a more than adequate score which was beaten by only three players, and now he was out on his own by 12 strokes. He returned to his hotel for lunch in the usual way, and I do not believe it entered the head of a single person present that they might be about to witness in the afternoon the most agonizing golfing spectacle any can remember to this day.

Things went instantly wrong. He timed his arrival for the start but found it postponed for a quarter of an hour owing to the immense crowd which had assembled to watch the triumphant formality of his final round. In his own words: 'Like a fool, I went and sat in a small tent all by myself. Lack of experience again.

Today I should go out and hit balls, go for a walk, anything bar sit and brood. Already I had been undermined by people congratulating me before I had won. The editor of one of the golf magazines seemed to think he had appointed himself my official manager and kept popping in and out of lunch telling me not to sign anything without consulting him when I had won. I had been humbled by golf too often. I sat and thought how anyone could take 82 in a championship, and anyone else could do 69, and there is the whole thing gone. Why, it was only a mile or two away, at Deal, that poor Abe Mitchell took 83 to George Duncan's 71 and lost 12 strokes and the championship in a single round.'

The start was a foretaste of what was to come, and I hardly like to write of it even now. His first drive was skied and his second with a 2-iron hit a lady, standing at cover point, on the knee. Through the green it is no exaggeration to say that a competent 12-handicap player would have given him a good game and, if he had not putted, considering the circumstances, miraculously, he might have taken 90.

There was much talk at the time of his having eaten something that disagreed with him or having failed to digest his modest lunch. The latter, I am sure, is true or he could not speak of it with such feeling to this day. 'I played in a cold sweat and wanted to be sick. I ought to have gone off and vomited in the nearest hedgerow, but I didn't, partly because I was too ashamed and partly because there aren't any hedgerows at St George's anyway. I could not get anything but fives. I could not get a 4 even at drive-and-pitch holes where all the week I had been looking for threes.'

At the long thirteenth—'another b—— 5 coming'—the course of the round, and with it, he now agrees, probably of his life, was changed. He holed a four-yard putt for 4. It broke the spell and he coasted home to win by five shots from Sid Brews, of South Africa. He missed a short putt on the last green but it did not matter now. A British player had won the championship at last and they carried him shoulder high off the green.

HENRY LONGHURST, Sunday Times

MOE NORMAN

I confess that I had never heard of Moe Norman until I came across this piece. He was obviously an extraordinary player and perhaps the ultimate embodiment of the distinction which Bobby Jones makes.

I think that a man may be a truly great golfer and not be a great tournament golfer; and I do not think that the customary implication, that a great golfer who fails to shine at formal competition lacks courage, is justified. Matters of physique and mere physical stamina have a profound effect, as do also personal inclination and taste. Then there is that curious and little understood factor of temperament, which is so convenient an explanation either of the successful tournamenteer or the unsuccessful one.

In any event, I maintain that golf and tournament golf are two different things.   R. T. JONES, *Down the Fairway*, 1927

❖

Watson was once waiting to hit a tee shot in a PGA Tour event when the conversation around the tee turned to the question of who were the pure swingers in the game. Watson wheeled round. 'Hey, I'll tell you about a guy who can hit it better than anybody. His name is Moe Norman, up in Canada.'

Trevino speaks with particular sensitivity about Norman. 'I don't know of any person that I've ever seen,' Trevino told his interviewer on a documentary about Norman for the Canadian Broadcasting Corporation, 'who could strike a golf ball like Moe Norman as far as hitting it solid, knowing where it's going, knowing the mechanics of the game, and knowing what he wanted to do with the golf ball. When you're talking about Moe Norman you're talking about a legend, and I'm talking about a living legend because the public doesn't know Moe Norman. Ask any golf professional, whether you're in Australia, the U.S. or Great Britain, and they say that's the Canadian guy that hits it so damn good, isn't it, and I say that's him. He's a legend with the professionals. The guy's a genius when it comes to playing the game of golf.'

Norman, who was born in 1929, uses an original style that appears to invent a new language of the game. He's an enigma. Talent such as his deserves center stage, but Moe has spent his life far from the golf world's spotlight. Still, there's no denying his talent. He's a prodigy.

Is golf cerebral? Norman has mastered it with instinct and imagination. Must technique conform to strict principles? Norman's method seems to defy convention while causing observers to wonder what's going on and to speculate about what makes his swing send the ball off so cleanly nearly every time. Is golf slow, a labored enterprise where the player has too much time to think? Norman plays it with the speed of a gazelle. He takes one look at his target and he swings. For him, this flick-of-a-switch speed neutralizes golf's psychological complexities. Most golfers take between ten and twenty seconds from address to swing completion. He takes three seconds. Norman calls himself 'the 747 of golf. One look at the target and I'm gone. Miss 'em quick. That's always been my theme song.' Officials at a tournament once asked him to zigzag his way down the fairways rather than walking straight down the middle; he was playing so quickly that they needed to slow him down.

Norman makes golf look as easy as tossing a coin. Whereas most golfers arrange themselves loosely over the ball before swinging, Norman appears rigid. He stands far from the ball, his arms stretched to the limit, his feet far apart in a wide stance. He places the clubhead a foot or so behind the ball rather than directly behind it. This simple, casually brilliant maneuver enables Norman to follow the golf dictum that one should take the club back from the ball long and low, combing the ground along the way. He's a foot back before he starts, and then swings his club straight back from the ball and through it, extending his arms and club farther, perhaps, than any golfer in the history of the game. His extension through the ball protects the clubhead from fluttering and accounts in part for his uncanny accuracy. He demonstrates his clubhead extension by putting a silver dollar on the ground thirty-seven inches behind the ball. The sole plate of his driver contacts the coin every time. Then he moves the coin twenty-two inches ahead of the ball and contacts it again every time. Every shot is straight, every swing a carbon copy of the one that came before. He's so accurate that he's worn a dark spot the size of a quarter in the middle of all his iron clubs.

There was the time Norman gave a clinic during a Canadian Professional Golfers' Association Seniors Championship in Winnipeg. He hit the flag three times. Another day, Norman was having breakfast in New Smyrna Beach, Florida, when fellow pro Ken Duggan asked him why he wasn't playing. 'I hit the flagpole the first three holes,' Norman said. 'Why go on? Can't do any better than that.' He shot sixty-nine during a Senior PGA Tour event in Vancouver while playing with Australian Ken Nagle; at four-under par through eleven holes, he then became upset on the twelfth hole when his shot spun back off the green into rough. 'It's one of the most amazing exhibitions of ball-striking I've ever seen,' Nagle said after the round. 'If he hadn't let that incident bug him, he would have shot sixty-three or sixty-four. But he kept repeating what had happened. He couldn't drop it.'

Some years ago I wandered into the pro shop at the Pinehurst #2 course in North Carolina. A fellow in the shop told me that Norman had just come through Pinehurst the week before on his way home from Florida, where he spends his winters, living cheaply.

'See the "A" in Advantage?' the fellow asked, pointing to a canvas backing hung fifteen feet from where a golfer would hit balls in the indoor teaching area for the Advantage Golf School. The word was printed on the canvas. 'Well, Moe Norman came in here last week and hit the bottom part of the "A" five times in a row.' That's nothing. Another time, he showed up at the Tomoka Oaks course in Daytona Beach to play four holes prior to a tournament. Up at 6:00 A.M., he was on the course at seven, hitting six balls from the tee. His friend, professional Ken Venning, showed up soon after and saw that three balls were touching. 'Am I seeing mushrooms?' Venning asked Norman. 'Or are those the shots you hit?' They were indeed Norman's shots, the three other balls were nearby.

Norman's skills have brought him some fifty pro tournament victories. He holds thirty-three course records. He turned fifty in 1979 and won the next seven Canadian Professional Golfers' Association senior championships. Yet he has never succeeded on the U.S. PGA Tour and refuses to try to qualify for the rich Senior PGA Tour. His prowess with a golf club begs the question: Why hasn't he succeeded on the world's tours? Why isn't his name all over the record books?

The answer lies in Norman's personality. He suffers from an inferiority complex so pervasive that it's kept him out of the mainstream of golf. George Knudson once said in a dinner line that Norman was second to none in ball-striking. He didn't know that Norman was behind him; Norman started to cry, so much respect did he have for Knudson's opinion, so little belief in himself. 'He is,' Knudson said, 'the most sensitive man I know.'

Paradoxically, Norman's overwhelming insecurity may be the source of his remarkable talent. I've come to believe in the twenty-five years that I've watched and gotten to know him that without his desire to get out of everyone's way, he might not be so speeded up. Without the speeding up that is the essential component of his golfing personality, he might have tried a more conventional way of hitting the ball. There may then have been no offbeat setup, no pure, instinctive swing, no mechanical precision. It's possible that his vulnerability—his shadow—has produced behavior that meshes exactly with the demands of the game itself but renders him unable to cope with the social and business part. I think that he knew without saying so that to try to adapt to the golf world would have meant the loss of his distinctive way of playing. Moe Norman likes hitting the golf ball too much to ever have considered such a sea change. Besides, Norman on the course moves to his own accelerated rhythm. He cannot be curbed, tamed, or slowed down.

I first encountered Norman in the early 1960s at the De Haviland Golf Centre in Toronto, where he worked as an assistant pro. His job was to entertain patrons with his maximum-efficiency swing.

Moe was unforgettable. He seemed a sorcerer come to the night to demonstrate what golf was all about. It wasn't about winning money or becoming a head professional at a fancy country club. It wasn't even about entering tournaments, or winning the Masters. It was about playing. It was about joy. It was a simple game best played in an almost naive way. He rang with delight while hitting balls.

'Let your body enjoy the shot,' Moe would tell the crowd. 'Let it enjoy the shot. That's the biggest word in golf, let.' And then he might define the game. 'Golf,' he said. 'It's hitting an object to a defined target area with the least amount of effort and an alert attitude of indifference.' Don't care so much about results, then. That's what I heard. Swing the club. Forget about the shot when it's over. Looking back on those days at De Haviland, I know what attracted me to Norman. It was the happiness he exuded when he stood over and hit the ball. It's the same quality I've seen since to varying degrees in other golfers. But it was and still is the main component of Moe's game.

While competing as a young amateur. Moe hitchhiked to events and slept in sand traps or on benches. He set pins in a Kitchener bowling alley to earn a few dollars and became the fastest pin-boy around, a whirling dervish of the lanes. Wherever he went, he drew attention. Golf was his one love. 'It became his life,' Marie says. 'He used to tell us that someday he would be the richest guy in the world because of golf.'

He certainly seemed on his way, winning both the 1955 and 1956 Canadian Amateur championships. After his first win, he was greeted by a group of fellow golfers who led him back to the course in a small motorcade. Moe sat in the corner at what was to be a celebration, too shy to say a word. Yet even as he turned inward, golf was propelling him forward. He was invited to the 1956 Masters. 'I was setting up pins one night when it was ten below,' he remembers, still awestruck, 'and when I got home, here's this invitation to the Masters. What a thrill.'

Surrounded by thousands of fans, Norman trembled on the first tee at the Augusta National Golf Club. 'I was shaking like a leaf. But before they could say "Moe Norman," bang, I hit it down the middle.' He conquered fear with speed, but he didn't complete the Masters. After the first round, Sam Snead gave him a lesson that so excited him that he hit 800 balls that evening. But by taking the lesson at a critical time, he betrayed his fundamental problem. He didn't believe in his own ability to play against such competition.

Four hours later, as darkness fell over the range, his hands were raw. He could hardly grip the club the next day because he had split a thumb; he withdrew after nine holes of the second round. It's been said that he walked off because play was too slow, or because he felt uncomfortable. Irv Lightstone, a Toronto professional who was with him, says Norman was in too much pain to continue.

Moe never felt comfortable in the United States. A pro told him that if he were going to play on the tour, he had better improve his grooming—his pants were often above his ankles, his toenails occasionally stuck through his shoes. And tour officials admonished him for his antics; he had hit balls off Coke bottles during the Los Angeles and New Orleans Opens. 'I was putting on a show,' he says, 'making the crowd laugh. But they told me this was big business, this was the tour of the world, that they didn't care how good I was, I had to tee it up in the normal way.'

After Moe played his ten United States events, he left the tour there for good, abandoning his childhood vision of getting rich from golf. He played the 1963 Canadian Open at the Scarboro Golf and Country Club, where Toronto Maple Leaf hockey star Teeder Kennedy told him that he could win. 'I'm a teaching pro,' Moe countered, 'not a playing pro.' Nevertheless, Moe played well enough to lie in second place after fifty-four holes, only three shots from the lead. But he started to think about what he would say if he won and became nervous. His agitation showed on the greens, where he three-putted six times. 'I blew that tournament,' he says. 'Tee to green I was comfortable. Not on the greens. Shaky like a leaf. No self-confidence. I didn't know how to get the ball in the hole. The Canadian Open was controlling me. I wasn't controlling it. It was eating me because I wanted to win so bad.' Moe hit the ball so well that it could make a person cry, but he was too frightened to compete. His form was immaculate, but he also embodied the golfing truth that errors in the swing often begin in the mind; his mind was shot through with fear.

To me, Moe remains the golfer at De Haviland. He's hitting golf balls into the night sky while showing a youngster that the game can be a source of delight. I don't know if I understand him any better now than I did thirty years ago. But that doesn't matter. What matters is that Moe Norman is still hitting golf balls, still wandering from course to course, still enjoying the meeting of the club-face against the ball. He swings, and he lives. It's impossible to think of Moe Norman without golf.

Henry David Thoreau wrote that the highest of the arts is to 'affect the quality of the day.' Moe has done this for me every time I have watched him.   LORNE RUBINSTEIN, *Links*, 1991

❖

ARNOLD PALMER

He first came to golf as a muscular young man who could not keep his shirttail in, who smoked a lot, perspired a lot and who hit the ball with all of the finesse of a dock worker lifting a crate of auto parts. Arnold Palmer did not play golf, we thought. He nailed up beams, reupholstered sofas, repaired air-conditioning units. Sure, he made birdies by the streaks in his eccentric way—driving through forests, lacing hooks around sharp corners, spewing wild slices over prodigious hills, and then, all hunched up and pigeon-toed, staring putts into the cups. But he made just as many bogeys in his stubborn way. Anyhow, a guy whose slacks are too long and turned up at the cuffs, who matches green shirts with orange sweaters, a guy who sweats so much, is not going to rush past the Gene Littlers, Ken Venturis and Dow Finsterwalds, the stylists, to fill the hero gap created by the further graying and balding of Ben Hogan and Sam Snead. This is what most of us believed around 1960, even after Palmer had won his second Masters, even after he had begun to drown everyone in money winnings. *This* was a suave new godlet of the fairways, a guy out of Latrobe Dry Goods?

We were, of course, as wrong about him as the break on a downhill six-footer, as wrong as his method seemed to us to be wrong: hit it hard, go find it, hit it hard again. We knew we were wrong one day when the bogeys suddenly went away. No one understood why or how, except that Palmer willed them to. And now he had become a winner like none we had ever known. He was a *nice* guy, of all things. He was honestly and naturally gracious, untemperamental, talkative, helpful and advising, unselfish of his time, marvelously good-humored; he had a special feeling for golf's history and he was honored by its traditions; and with all of this he remained the gut fighter we insisted he be, a man so willing to accept the agonies of pressure and the burdens of fame that for a few years we absolutely forgot that anyone else played the game he was dominating and changing.

He actually started *being* Arnold Palmer in that summer of 1960, a stupidly short time ago it seems. He became the Arnie of whooha, go-get-'em Arnie on a searingly hot afternoon in Denver when, during the last round of the U.S. Open, he exploded from seven strokes and fourteen players behind to win. Two months earlier he had finished birdie-birdie on national television to win the Masters and now he had created another miracle—again on national television.

Much has been written into the lore of golf of how it was that day, of the epic 65 he shot in the final round at Cherry Hills, of the day that really made him, but not by anyone who had lunched with him, kidded him, and then happily marched inside the gallery ropes with him, scurrying after Cokes, furnishing cigarettes, and hoping to put him at ease.

During lunch in a quiet corner of the Cherry Hills locker room before that round, Arnold was cheerful and joking as he ate a hamburger, drank iced tea, and made small talk with a couple of other players, Bob Rosburg and Ken Venturi, a writer named Bob Drum, and myself. He talked of no one else who might win. All he seemed concerned about was Cherry Hills's 1st hole, a comparatively short, downhill, downwind, par four. It bugged him. He thought he could drive the green, but in three previous rounds he had not done it.

'It really makes me hot,' he said. 'A man ought to drive that green.'

'Why not?' I said. 'It's only three hundred and forty-six yards through a ditch and a lot of high grass.'

'If I drive that green I might shoot a hell of a score,' he said. 'I might even shoot a sixty-five if I get started good. What'll that bring?'

'About seventh place. You're too far back.'

'Well, that would be two-eighty,' Arnold said. 'Doesn't two-eighty always win the Open?'

'Yeah,' I said. 'When Hogan shoots it.'

Arnold laughed and walked out to the first tee.

For a while I loitered around the big clubhouse waiting for the leaders to go out, as a good journalist should. In the process of milling around, however, I overheard a couple of fans talking about an amazing thing they had just seen. Palmer, they said, had driven the first green. Just killed a low one that hung up there straight toward the mountains and then burned its way through the USGA trash and onto the putting surface. Got a two-putt birdie.

I smiled to myself and walked out onto the veranda and began edging my way through the spectators toward the 1st tee where the leader, Mike Souchak, would be going off presently. But about that time a pretty good roar came up from down on the front nine, and seconds later, a man sprinted by panting the news that Palmer was three under through three.

'Drove the first, chipped in on two and hit it stiff on three,' he said, pulling away and darting off to join Arnie's Army. Like the spectator and a few thousand others who got the same notion at the same time, I tried to break all records for the Cherry Hills Clubhouse-to-Fourth Fairway Dash. We got there just in time to see Arnold hole his fourth straight birdie.

Wringing wet and perishing from thirst, I staggered toward the fifth tee, stopping to grab a Coke at a concession stand. I ducked under the ropes as an armband permitted and stood there puffing but excited.

Arnold came in briskly, squinted down the fairway and walked over. He took the Coke out of my hand, the cigarettes out of my shirt pocket and broke into a smile.

'Fancy seeing you here,' he said. 'Who's winning the Open?'

Palmer birdied two more holes through the 7th to go an incredible six under, working on an incorrigible twenty-nine out. But he bogeyed the 8th and had to settle for a 30. Even so, the challengers were falling all around him like wounded soldiers, and their crowds were bolting toward him, and the title would be his. Everything would be his now.

Later on, somewhere on the back nine holes, I remember sizing up a leader board with him and saying, 'You've got it. They're all taking gas.'

'Aw, maybe,' he said, quietly. 'But damn it, I wanted that twenty-nine.'

There have been other major victories, as we know, and scores of lesser ones, and precisely because of him the professional tour has tripled, quadrupled in prize money. He has become, they say, something immeasurable in champions, something more than life-size, even though he has turned into his forties, the hip hurts, and a lot of other big ones have slipped away.

This is true, I think. He is the most immeasurable of all golf champions. But this is not entirely true because of all that he has won, or because of that mysterious fury with which he has managed to rally himself. It is partly because of the nobility he has brought to losing. And more than anything, it is true because of the pure, unmixed joy he has brought to trying.

DAN JENKINS, The Dogged Victims of Inexorable Fate, 1970

JACK NICKLAUS

Mention was made in an article from Pebble Beach of Nicklaus at practice, and of how drive after drive flew down the precise line to a caddie in the remote distance, with an unwavering regard for the shortest distance between two points. The impression was not so much of a ball driven by a swing, which certainly is true of Palmer for all that he hits so hard, but rather as if it had been propelled by a hammer blow from some mechanical device.

Nicklaus is so strong that he does not require a full swing, and at the top of it the club never reaches the horizontal. The backlift is simple with no break in the wrists until the widest possible arc away from the ball has been achieved. Then, from a stance which makes clearance of the left side after impact as easy as possible, and an anchorage so solid that it might well be cast in concrete, so massive are his legs, Nicklaus lets fly. There is no suggestion of a slash. Considering the force involved, the whole method is remarkably balanced and controlled.

The same strength is revealed in his long irons, and, like Palmer, Snead and J. Hebert, among the few leading golfers who do, Nicklaus carries a number one. In one round he was green high in two at the eighteenth, well over 500 yards, using this club for the second shot in still air.

It was interesting to examine his clubs, for their comparative lightness, and to see how low on the blade the beautifully compact mark of impact was, compared to that made by most other golfers. This was evidence of an abnormal precision and purity of striking but, all these considerations of strength aside, much more is required to have reached the peak of achievement already attained by Nicklaus.

On the second hole after lunch in the final, his ball lay on sandy grass. A wide bunker was between him and the pin, which was placed just beyond, and yet, with a swing so delicate in rhythm that it concealed the firmness beneath, he played the shot perfectly to within two feet of the hole. He has a remarkable sensitivity of touch in the short game, and his pitching with wedge or short irons is wonderfully exact. Time after time he would finish within a few feet of the hole, invariably leaving himself with a putt for a birdie. This standard is unapproached by the British, which is one of the principal causes for their failure in competition against Americans. Nicklaus always attacks with this shot to the pin, and with the putt that follows. Never has one seen a golfer whose whole appearance on the greens suggested a greater degree of determined concentration.

His putting method is similar to Palmer's except that the knees are not locked inwards to the same extent. The body is crouched over the ball, with the head well down; and the stroke is the usual crisp American tap, the right hand pushing the club through low along the line. The sight of Nicklaus on the green, the chunky body absolutely motionless save for the quick turns of the head as he checks the line with those clear blue eyes of his, the ash-blond hair peeping from beneath his white cap, and the set lines of his face, remain an enduring picture . . .

PAT WARD-THOMAS, *The Long Green Fairway*, 1966

SAM SNEAD AND ARNOLD PALMER

Though they no longer are likely to win, you wouldn't know it from their charismas. Snead, with his rakishly tilted panama and slightly pushed-in face—a face that has known both battle and merriment—swaggers around the practice tee like the Sheriff of Golf Country, testing a locked door here, hanging a parking ticket there. On the course, he remains a golfer one has to call beautiful, from the cushioned roll of his shoulders as he strokes the ball to the padding, panther-like tread with which he follows it down the center of the fairway, his chin tucked down while he thinks apparently rueful thoughts. He is one of the great inward golfers, those who wrap the dazzling difficulty of the game in an impassive, effortless flow of movement. When, on the green, he stands beside his ball, faces the hole, and performs the curious obeisance of his 'sidewinder' putting stroke, no one laughs.

And Palmer, he of the unsound swing, a hurried slash that ends as though he is snatching back something hot from a fire, remains the monumental outward golfer, who invites us into the game to share with him its heady turmoil, its call for constant courage. Every inch an agonist, Palmer still hitches his pants as he mounts the green, still strides between the wings of his army like Hector on his way to yet more problematical heroism. Age has thickened him, made him look almost muscle-bound, and has grizzled his thin, untidy hair; but his deportment more than ever expresses vitality, a love of life and of the game that rebounds to him, from the multitudes, as fervent gratitude. Like us golfing commoners, he risks looking bad for the sake of some fun.

Of the younger players, only Lanny Wadkins communicates Palmer's reckless determination, and only Fuzzy Zoeller has the captivating blitheness of a Jimmy Demaret or a Lee Trevino. The Masters, with its clubby lifetime qualification for previous winners, serves as an annual exhibit of Old Masters, wherein one can see the difference between the reigning, college-bred pros, with their even teeth, on-camera poise, and abstemious air, and the older crowd, who came up from caddie sheds, drove themselves in cars along the dusty miles of the Tour, and hustled bets with the rich to make ends meet. Golf expresses the man, as every weekend foursome knows; amid the mannerly lads who dominate the money list, Palmer and Snead loom as men.   JOHN UPDIKE, *At the Masters*

AS A PARADE OF LOVELY GOLFERS, NO TWO ALIKE

Charles Coody, big-beaked bird. Billy Casper, once the king of touch, now sporting the bushy white sideburns of a turn-of-the-century railroad conductor, still able to pop them up from a sand-trap and sink the putt. Trevino, so broad across he looks like a reflection in a funhouse mirror, a model of delicacy around the greens and a model of affable temperament everywhere. Player, varying his normal black outfit with white slacks, his bearing so full of fight and muscle he seems to be restraining himself from breaking into a run. Nicklaus, Athlete of the Decade, still golden but almost gaunt and faintly grim, as he feels a crown evaporating from his head. Gay Brewer, heavy in the face and above the belt, nevertheless uncorking a string-straight mid-iron to within nine inches of the long seventh hole in the par-three tournament. Miller Barber, Truman Capote's double, punching and putting his way to last year's best round, a storm-split 64 in two installments. Bobby Clampett, looking too young and thin to be out there. Andy Bean, looking too big to be out there, and with his perennially puzzled expression seeming to be searching for a game more his size. Hubert Green, with a hunched flicky swing that would make a high-school golf coach scream. Tom Weiskopf, the handsome embodiment of pained near-perfection. Hale Irwin, the picture-book golfer with the face of a Ph.D. candidate. Johnny Miller, looking heavier than we remember him, patiently knocking them out on the practice tee, wondering where the lightning went. Ben Crenshaw, the smiling Huck Finn, and Tom Watson, the more pensive Tom Sawyer, who, while the other boys were whitewashing fences, has become, politely but firmly, the best golfer in the world.

And many other redoubtable young men. Seeing them up close, in the dining room or on the clubhouse veranda, one is struck by how young and in many cases how slight they seem, with their pert and telegenic little wives—boys, really, anxious to be polite and to please even the bores and boors that collect in the interstices of all well-publicized events. Only when one sees them at a distance, as they walk alone or chatting in twos down the great green emptiness of the fairway, does one sense that each youth is the pinnacle of a buried pyramid of effort and investment, of prior competition from pre-teen level up, of immense and it must be at times burdensome accumulated hopes of parents, teachers, backers. And with none of the group hypnosis and exhilaration of team play to relieve them. And with the difference between success and failure so feather-fine.   Ibid.

THREE EXTRAORDINARY LADIES

In relation to the numbers playing, ladies' golf is undoubtedly much underrepresented in this collection. This is certainly no criticism of the quality of their play, which is today extremely high, but it does mirror the rather low-key presentation of the ladies' game, especially in the UK.

The relationship between the men and the women as regards skill and strength seems to be very much the same for golf as it is for tennis. Yet tennis produces a string of famous women champions who command a level of attention from media and public—and not least from sponsors—which has no equivalent in golf. Why this should be the case is a mystery—at least to me. Presumably the answer lies in something deep in the structure of the game and its institutions, which has resulted in the complete separation of the sexes as far as tournaments and competitions are concerned; but whatever it is, there is no doubt that ladies' golf has lost out as a result.

JOYCE WETHERED

The following tribute is an indication of the special standing of Miss Wethered (later Lady Heathcoat-Amory) and the extraordinary quality of her play.

Ordinarily I would never take advantage of a friendly round of golf by making the play of a person, kind enough to go around with me, the subject of an article. I realize that everyone likes to play occasionally a round of golf when reputations can be forgotten, with nothing more at stake than the outcome of the match and a little friendly bantering afterwards.

Just before the British Amateur championship at St Andrews, Miss Joyce Wethered allowed herself to be led away from her favorite trout stream in order to play eighteen holes of golf over the Old Course in company with her brother, Roger, Dale Bourne, then recently crowned English champion, and myself. At the time, I fully appreciated that Miss Wethered had not had a golf club in her hand for over a fortnight, and I certainly should have made no mention of the game had she not played so superbly.

We started out by arranging a four-ball match—Roger and Dale against Miss Wethered and myself—on a best and worst ball basis. I don't know why we didn't play an ordinary four-ball match, unless we fancied that the lady would be the weakest member of the four, and that in a best-ball match her ball would not count for very much. If any of us had any such idea at the start of the match, it is now quite immaterial, for there is not the slightest chance that we should admit it.

We played the Old Course from the very back, or the championship tees, and with a slight breeze blowing off the sea. Miss Wethered holed only one putt of more than five feet, took three putts rather half-hearted from four yards at the seventeenth after the match was over, and yet she went round St Andrews in 75. She did not miss one shot; she did not even half miss one shot; and when we finished, I could not help saying that I had never played golf with anyone, man or woman, amateur or professional, who made me feel so utterly outclassed.

It was not so much the score she made as the way she made it. Diegel, Hagen, Smith, Von Elm and several other male experts would likely have made a better score, but one would all the while have been expecting them to miss shots. It was impossible to expect that Miss Wethered would ever miss a shot—and she never did.

To describe her manner of playing is almost impossible. She stands quite close to the ball, she places the club once behind, takes one look toward the objective, and strikes. Her swing is not long, surprisingly short, indeed, when one considers the power she develops, but it is rhythmical in the last degree. She makes ample use of her wrists, and her left arm within the hitting area is firm and active. This, I think, distinguishes her swing from that of any other woman golfer, and it is the one thing that makes her the player she is.

Men are always interested in the distance which a first class woman player can attain. Miss Wethered, of course, is not as long with any club as the good male player. Throughout the round, I found that when I hit a good one I was out in front by about twenty yards—by not so much when I failed to connect. It was surprising though how often on a fine championship course fine iron play by the lady could make up the difference. I kept no actual count, but I am certain that her ball was the nearest to the hole more often than any of the other three.

I have no hesitancy in saying that, accounting for the unavoidable handicap of a woman's lesser physical strength, she is the finest golfer I have ever seen.   R. T. JONES, *American Golfer*, August 1930

MISS MINOPRIO

No one would pretend that Miss Minoprio was in the same class as Miss Wethered or indeed dozens of other ladies as a player, but she nevertheless earned a special place in history by her eccentricity and, no doubt, did a good deal to help relax the conventional view of the appropriate dress for the game.

It was ten o'clock on a lovely October morning and the scene was Westward Ho! The flat expanse between the golf club-house and the sea, which they call the 'Burrows', was shimmering in the sunshine, and red sails glided along above the level of the sandhills as the sailing barges and fishing boats made their way into the little harbour at Appledore. The world was at peace.

On the first tee down below the club-house a small group of people were waiting. Many were women, for the English women's championship was just beginning, but among them was a goodly sprinkling of men. Of the men, those that weren't caddies were golf journalists.

Golf journalists on the first tee at ten in the morning? Yes, indeed. Every one of them. And what had lured them forth at this unaccustomed hour? Why, the rumour had gone round the village that at ten o'clock that day a lady intended to play golf in trousers.

One or two even went so far as to suggest that not only did she play in trousers but that she only used one club. This was ruled out as an unworthy attempt to paint the lily. Trousers, yes; or one club, yes. But trousers and one club—come, come, sir!

No one you meet has ever seen a ghost; on the other hand, there's no one who doesn't know someone who has. So it was with the mysterious lady. No one had seen her, but every one had it first-hand from someone who had.

Ten o'clock came, and no apparition. The name was called once. No reply. It was called again. No reply. The know-alls wagged their heads with a chorus of 'I told you so', and were retiring to the

club-house for refreshment, when along the little lane that cross-
es the links a couple of hundred yards from the first tee there
appeared a big yellow motor car.

The car stopped, and out into the headlines stepped Miss Gloria
Minoprio.

The deserters hastily retraced their steps from the bar, while
among the ladies in waiting arose a clucking and fluttering as of
an agitated flock of Leghorn pullets.

'My *dear*, do you see what I see? . . .' 'What a figure! . . .' 'What
trousers!'

'Well, *really!*' cried the Ladies' Golf Union.

'Good God!' said the journalists.

Meanwhile, the object of their astonishment made her way com-
posedly and with what dignity her costume would permit across
to the waiting crowd.

She was clad from head to foot in dark blue, and, yes, she wore
trousers. Close-fitting, exquisitely tailored trousers, very tightly cut,
especially—er—behind. She wore them, as did our grandfathers,
with straps beneath the insteps of her blue suède shoes. A neat
blue jacket and a little blue turban completed the streamline.

A slim, graceful girl, with delicate, sensitive features and figure
divine. She had bumps, to quote Mr Damon Runyan's rudely graph-
ic description, where a doll is entitled to have bumps. Only one
thing marred the picture. On her cheeks should have glowed the
rosy bloom of youth and health. Instead, they were heavily, almost
grotesquely, powdered in white. She might have been wearing a
white mask.

With her was a young caddie carrying, rather sheepishly, a scarlet spare jacket, a ball bag, and—not one club, but two. But rumour
had spoken truth, and she only used one of these clubs. The other
was a spare in case of accident.

She said 'How do you do?' almost inaudibly to her opponent. At
the end of the match she said 'Thank you.' So far as I am aware
that was all she did say.

Tapping the ground with her solitary cleek to show the caddie
where she wished him to tee the ball, she prepared to play her
opening stroke. It must have been something of an ordeal. If so,
she certainly showed no sign of it.

She had, it turned out, a careful, precise style of play that might
have been learned studiously from the text-book. Nothing very

dashing about it, no undignified vigour, but quite efficient. She did not hit the ball very far—no woman does with an iron club—but she hit it for the most part nice and straight.

That morning the champions played in solitude, their supporters lured away by magnetic Minoprio. I forget the name of her opponent, but there was no doubt as to which was the more nervous of the two. The prospect of losing to a lady in fancy dress using only one club is enough to shake the stoutest heart. Might take a lifetime to live down.

Recovering from their initial shock, those of the quickly gathering gallery who were interested in the technique of golf settled down to assess Miss Minoprio's capabilities with her solitary club. They proved to be considerable. Her long game was steady and, though the long shaft of her iron made her look rather clumsy, her putting was at least up to the average usually seen in a women's championship. Her approaches, low along the ground, were quite effective.

But the time came, inevitably, when she was faced with strokes beyond the capacity of Bobby Jones, Cotton, or the devil himself, to execute with a straight-faced iron. She could not loft the ball, except in a full shot; she could impart no 'stop' or back spin; she could get no distance from anything but a smooth, clipped lie. To lob the ball over an intervening hazard was beyond her.

There was much speculation as to what would occur when she got into a bunker. The truth is that it is quite simple to remove the ball from a bunker with any kind of club if the sand is soft and loose. Hit hard, three or four inches behind it, and the deed is done. So in the fine seaside sand of Westward Ho! Miss Minoprio performed with no little distinction, and some who had come to mock remained to marvel. But on firm or rain-sodden sand, or, indeed, on any hard surface, she was pathetically powerless.

That her average score would have been reduced by anything from half-a-dozen to ten shots in a round by the use of a normal set of clubs, no reasonable critic could doubt. She went through the motions well enough, but the instrument she used was too ill adapted for the purpose. She lost her match by, I think, five and four, and I was able to telephone to the *Evening Standard* what I believe to be almost the only Latin tag to find its way into the sporting pages of that journal—'Sic transit Gloria Monday.' (I repeated it shamelessly for five years with only one variation. One year she defeated a young girl who was so nervous that she could scarcely focus the ball. So when she was beaten next day the tag became 'Sic transit Gloria Tuesday.')

The yellow car had been driven across the 'burrows' and was waiting near by. She stepped in and was whisked away, not to appear in public again until the next women's championship, and the company settled down to debate her reason for imposing upon herself the ludicrous handicap of playing with one club.

One school of thought held that she was doing it for publicity. Certainly her unusual attire lent weight to that opinion. If so she certainly succeeded, for her name and picture have featured in almost every newspaper every time she has appeared in a championship. But those who seek publicity like inevitably to bask in it when achieved. Miss Minoprio, so far as I know, has never entered a golf club-house during a championship; has never played in a tournament other than the two championships; has never made friends with other golfers.

And again, why the extraordinary outfit—admirable though it may be for golfing comfort? And why the mask-like countenance? Here a very strong school exists which holds that she plays golf while temporarily hypnotized, or entranced, either by auto-suggestion or by a friend. That is possible. And if she is not in a state of semi-hypnosis, that at least is as good a description as any of her appearance and demeanour.

She spent some months studying yoga in India (and incidentally she is a conjurer of the highest order, though the point is hardly relevant). After her début at Westward Ho! she wrote to tell me she had bought thirty copies of the *Tatler*, in which I had written about her, to distribute to her friends. We exchanged three or four letters. Later, while walking round in a championship, I introduced myself as her correspondent. She blinked with surprise at being spoken to. She had the vacant, far-away look of Lady Macbeth walking in her sleep. She seemed scarcely to understand what I was trying to say. 'Oh, yes . . . yes,' she said, looking vaguely into the distance over my shoulder. I faded away.

The Ladies' Golf Union, aghast at her first appearance, issued a proclamation that they 'deplored any departure from the traditional costume of the game', but the last laugh was against them. Nearly half the field in women's championships today turn out in trousers.

But none of them fit like Gloria Minoprio's.   HENRY LONGHURST

BABE ZAHARIAS

With the same flair for showmanship as Walter Hagen (and much the same ruthless competitiveness), Babe Zaharias was the real driving force behind the founding of the US LPGA. Just as Vardon and Taylor forced their fellow professionals to raise the level of their game to match them, Zaharias made women players realize that there was no physical reason why they could not hit the ball as far as most men. Perhaps more than anyone, she determined the standards of excellence which the modern professional women's game has achieved.

It was her versatility, her excellence in so many sports—basketball, hurdles, discus, high jump, javelin, swimming, tennis, bowling— that made Babe Didrikson Zaharias singular, a woman athlete without peer. But it was golf that brought her lasting fame—and most of the million dollars she earned in her lifetime. She was not the best woman golfer of her era. That distinction belongs to a splendid Briton, Joyce Wethered, who won the British amateur four times in the '20s, turned professional for a brief, profitable U.S. tour in the mid-'30s, then wed Lord Heathcoat-Amory and retired behind the genteel hedges of an estate in Devon.

But Babe Zaharias created big-time women's golf. She launched it as a legitimate sport and brought gusts of freshness and fun to a game too often grim. She joked and clowned and had a rapport with fans that is rare. She had the ability to be cocky with charm, and the galleries loved her. Her booming power game lowered scores and forced others to imitate her. And had it not been for her death—in 1956 when she was just 45—she would reign as the game's dowager queen.

When Babe won her first golf tournament—the 1935 Texas women's championship—Grantland Rice took note in his celebrated tin-ear doggerel:

> From the high jump of Olympic fame,
> The hurdles and the rest,
> The javelin that flashed its flame
> On by the record test—
> The Texas Babe now shifts the scene
> Where slashing drives are far
> Where spoon shots find the distant green
> To break the back of par.

But Babe, then 23, had had to crack something much more testing than the back of par—Texas golf society. She had no pedigree, coming as she did from a dead-end neighborhood in Beaumont, no money and not much social grace. Her gold medals from the 1932 Olympics counted for little among the country-club set, and her fame had already faded. There was only her golf game, at that point strong but scarcely smooth. When she entered the Texas event, a member of the Texas Women's Golf Association named Peggy Chandler declared, 'We really don't need any truck drivers' daughters in our tournament.'

Several women withdrew from a driving contest that preceded the tournament, implying that Babe was too manly for them to compete against. Babe purposely dubbed drive after drive with an exaggerated girlish swing—except one that she hit 250 yards to win.

In the tournament she overwhelmed her first three opponents and won her semifinal match in the rain on the 18th hole with a 20-foot putt that spurted water across the green. In the final she met none other than Peggy Chandler, and the match was B-movie material: scruffy poor girl vs. snobbish rich girl. They played 36 holes before a large gallery. By the 26th hole Peggy Chandler was 3 up, but Babe rallied, and on the 30th hole she drew even. On the 34th hole, a long par-5, Peggy Chandler put her third shot on the green close to the cup and a birdie seemed certain. Babe's drive had been a wild 250-yarder into a ditch. Her second, a three-iron, carried over the green, the ball coming to rest in a rut containing an inch of water. Her next shot was pure penny-dreadful heroics, a pitch that rolled into the hole for an eagle 3. The golfers halved the 35th and Babe won the 36th to take the match 2 up.

Now, if the B-movie scenario were truly followed, Babe would have been welcomed into the perfumed society of Texas golf. Instead, the U.S. Golf Association, acting on information presumably supplied by some of the Texas women, ruled that it was in 'the best interest of the game' that Babe Didrikson be barred from amateur golf, that she was a professional. This meant Babe was eligible to compete in only one tournament, the Western Open. There was no other event in the world for women pros (the U.S. Women's Open was first played in 1946). Babe turned for help to R.L. and Bertha Bowen of Fort Worth, friends she had made during the Texas tournament. R.L. was president of Community Public Service, an electric light and power company. Bertha was one of the group that ran women's golf in the state, and she had been appalled at the clawing Babe had received. The Bowens invited Babe to their home. They called a lawyer; they contacted the USGA; it was no use. About the only thing they could do was start their own tournament for Babe and pros like her, the Texas Women's Open. Bertha recalls, 'I was furious that Babe had been cut off. I was criticized by some of my friends for befriending Babe. They'd ask, "Why are you fooling around with *that* girl?" '

Babe had rough edges then and her association with the Bowens supplied some social polish. 'She was so poor it was pitiful,' Bertha says. 'One night we were invited to a formal party and we asked Babe to come along. She hemmed and hawed because she didn't have any clothes. Well, we got her an evening dress and she took one look at it and said, "I'm not going to wear that naked thing." She was very modest. We had to chase her all over the house before we could get the dress on her. We finally cornered her in the kitchen and forced her into it.'

Most members of the Texas Women's Golf Association remained cool toward the Bowens' protégée; the ladies kept telling Bertha that Babe really must wear a girdle when she played golf. Babe put one on, played a round and returned in a frenzy to the Bowens. 'I heard the car come screeching in the driveway,' says Bertha, 'and Babe came tearing into the house. She was yelling, "Goddam! I'm chokin' to death!" As far as I know, she never put on a girdle again.' (Nonetheless, Babe's favorite wisecrack to galleries was: 'When I wanta really blast one, I just loosen my girdle and let 'er fly!')

The metamorphosis in Babe's appearance was considerable, but there seemed no solution to the problem of making a living from sport. With only two tournaments open to her, Babe had no choice but to go on the road. In the summer of 1935 she toured with Gene Sarazen. They played 18-hole exhibitions, Babe receiving $150 for each, a solid sum in those Depression times, when many teachers got less than $1,300 a year. Sarazen, now 73, remembers, 'She was still a big draw because of the Olympics. One odd thing, she always wanted to be paid in one-dollar bills. I'd go to write her a check and she'd say, "No, Squire, make it in ones." Then she'd stack them up and mail them to her bank in Beaumont.'

In 1938 Babe married George Zaharias, then a prosperous wrestler who made as much as $15,000 a night, and for a considerable time that ended her financial concerns. By 1940 she felt able to publicly renounce professional golf, and she settled down to the prescribed three-year purification period that would enable her to regain her amateur status. On Jan. 21, 1943 she celebrated becoming an amateur by winning the California state championship.

Babe dominated amateur golf as no woman ever has, winning 17 significant tournaments in a row during 1946 and 1947. Among them was the U.S. Women's Amateur (the final of which she won by a record 11 and 9), the All-American at Tam O'Shanter and the Titleholders, a mini-Masters, in which she overcame a 10-stroke deficit to win by five.

But her most notable victory was in the 1947 British amateur at Gullane, Scotland. No American had ever won this championship. Though it was Babe's first trip to Europe, she could not have been more at home. Each day she strolled the cobbled streets of Gullane on her way to the course, nodding and hollering Texas howdies. Gullane was a rough and lively links, lashed by wind and inhabited by sheep. When Babe played her practice rounds, the club arranged for a man to accompany her and sweep the greens of sheep droppings before she putted. She was her usual exuberant self, on one occasion doing a highland fling on the clubhouse lawn.

Babe's flamboyance and ungirdled power caught the imagination of Great Britain. She had won her six matches with ease, and the *Manchester Guardian* enthused: 'Surely no woman golfer has accomplished in a championship what Mrs Zaharias has achieved in this one. . . . She has combined in a remarkable way immense length with accuracy, reaching with a number-five iron holes at which others are content to be short with a wood. She is a crushing and heart-breaking opponent.'

Babe came home a heroine. A month later she won the Broadmoor invitational—her 17th straight. Then, in August, she once again decided to become a professional, signing a contract with Fred Corcoran, who had promoted the men's professional tour for 11 years.

There was not much money to be made from pro golf then; Babe earned most of her income playing exhibitions. By now she was receiving $600 for an appearance, while Ben Hogan and Sam Snead were getting $500. Corcoran also booked her into baseball parks—for $500—where she would hit trick golf shots and play third base during batting practice.

It was not enough for Babe. In the winter of 1948 she and George met with Corcoran and Patty Berg to found the Ladies Professional Golf Association. L. B. Icely, the president of Wilson Sporting Goods (which paid Babe $8,000 a year to promote its products), put up the money to start the tour in 1949. The original LPGA had six members, with Patty Berg as president. Total prize money the first year was about $15,000; Babe earned the most, $4,300, and she played in eight of the nine events. The following year she won six tournaments to finish No. 1 on the money list again, with $13,450.

As the tour grew richer, an intense rivalry developed between Babe and Louise Suggs. Suggs was stoical, very serious, colorless, humorless. She was a good golfer and her record in the early '50s almost equaled Babe's. Yet Suggs never got the kind of coverage Babe did. When Suggs won, the headlines were as likely to read. BABE LOSES as LOUISE WINS. When Babe celebrated her last birthday in 1956 in a Galveston hospital, every member of the LPGA sent her flowers except Louise. 'I didn't because I'm not going to be a hypocrite,' she explained.

A couple of women who competed in the early days of the LPGA deny that Babe Zaharias was the key to the organization's success, but even they acknowledge that her personality and talent attracted interest to tour events. Patty Berg says, 'Our sport grew because of Babe. She had so much flair, color and showmanship, we needed her. Her power astonished galleries.'

Babe once hit a drive 408 yards—according to *Time*—but there was much more to her game than strength. She was an intelligent golfer. She had graceful, tapered hands and her short game was soft and certain. Even her stance had an athletic buoyancy.

She was ferociously competitive, even with her closest friends. Once, during a practice round, she lent Betty Dodd a driver and another time she gave Peggy Kirk Bell an 11-iron. Both immediately used the clubs to splendid advantage; in both cases, after a few holes, Babe took the club back.

Babe would attempt to psych her opponents in every possible way. She would hit a five-iron, quickly stuff it in the bag and tell everyone it was a seven. She would stride onto the putting green before a tournament and shout, 'Are you girls practicin' to come in second?'

During her professional years Babe frequently was offered and got appearance money for participating in events. This is not done in the U.S.—though common in other countries—and a number of players on the LPGA tour resented it. In the early '50s golfer Betty Hicks publicly criticized Babe on this account. Babe, then president of the LPGA, called a meeting of the members and said: 'Now let me tell you girls something. You know when there's a star like in show business? The star has her name in lights, right? Well, I happen to be the star of this show and all the rest of you are in the chorus. People come to see the star and the star gets the money.'

She was the star. For as long as she was able.

WILLIAM OSCAR JOHNSON and NANCY WILLIAMSON, *Sports Illustrated*,
20 October 1975

In 1953 she was diagnosed as having cancer and underwent a major operation. Within four months she was back competing on the Tour, and in the following year she won five tournaments to finish second in the money order. Two years later, at the age of only 42, she died.

TOURNAMENTS AND
MATCHES

❖

*A*lthough *tournaments and championships were always important from the point of view of prestige, in the early days they were few in number and low in prize-money. Peter Lewis, the director of the British Golf Museum at St Andrews, has shown that in none of the four years 1901–4 were there more than five tournaments offering total purses over £100 and that in 1903, for example, there were only two—the Open and the News of the World.*

The only way, therefore, that the leading professionals could openly exploit their skills financially was through exhibitions and challenge matches. These sometimes involved enormously punishing travel schedules and did undoubtedly impose a severe strain on the players. When Bobby Jones saw Vardon and Ray at East Lake in 1913 they were on a three-month tour which meant 35,000 miles of travel by train and boat and allowed them an average of only one night a week in a hotel.

After the formation of the PGA in 1901, the number of tournaments was gradually increased, and although the exhibition circuit continued, the challenge matches died out. Nowadays, of course, they are virtually unknown. No leading professional would need or wish to risk his reputation over thirty-six or seventy-two holes. Apart from the Ryder Cup, the World Match Play Tournament at Wentworth, and the Dunhill Cup, the nearest that we now get are those occasions when the last round of a tournament develops into a man-to-man contest like the wonderful battle between Nicklaus and Watson at Turnberry in 1977. However the situation arises, there is no doubt that a keenly fought match is the most exciting form of the game for the spectator, and that it is the drama of match play that has provoked some of the most memorable writing.

ST ANDREWS v. MUSSELBURGH
In 1849 a challenge match was arranged between Allan Robertson and Tom Morris of St Andrews and the Dunn brothers of Musselburgh. It was to be played over three courses, two rounds each. At Musselburgh the brothers won by no less than 13 and 12, but this was reversed on their home course by the

[119]

St Andrews men. They therefore came to North Berwick with all to play for—and a stake of £400.

This enormous sum no doubt goes some way towards accounting for the final loss of nerve by the Dunns, just as the amount of betting by the spectators—who at one time laid odds of 20–1, against St Andrews—accounts for the size and excitement of the crowd. The description of this final is so fresh and vivid that the thrill of that day still comes bubbling through.

But the finest foursome of all that I remember was that between Allan and Tom against the two Dunns in the final at North Berwick. It created intense interest in the golfing world of that day, and crowds flocked to North Berwick to see it. I crossed over from Leven (Fife) with my brother James, and remember it well. When I awoke at five o'clock the rain was pouring, and I got up and told my brother so, and that it would be useless to go. However, in a short time afterwards, he came to my bedroom and said 'Man, Tom, I see a wee glint of blue sky! I think we should gang.'

'All right!' I said, 'I'm up.' And in due time we arrived at North Berwick.

On meeting Allan, I said I had come to see him win. He replied that he hoped so; but he had a dejected look about him, and I got the impression that he was doubtful of the result. The match was one of thirty-six holes, which required five or seven rounds (I forget which) of the North Berwick Links at that time, and one hole more.

The match started amidst the greatest enthusiasm. The weather had cleared up, but the wind blew pretty strong from the southwest. Each party had its own tail of supporters, those for the Musselburgh men predominating—for which, of course, the proximity of that place to North Berwick might account. They were led by Gourlay, the ball-maker. I never saw a match where such vehement party spirit was displayed. So great was the keenness and the anxiety to see whose ball had the best lie, that no sooner were the shots played than off the whole crowd ran, helter-skelter; and as one or the other lay best, so demonstrations were made by each party.

Sir David Baird was umpire, and a splendid one he made. He was very tall and so commanded a good view of the field; but it took all his firmness to keep even tolerable order.

The early part of the match went greatly in favour of the Dunns, whose play was magnificent. Their driving, in fact, completely over-powered their opponents. They went sweeping over hazards which the St Andrews men had to play short of. At lunch time the Dunns were four up, and long odds were offered on them.

On resuming the match, the advantage went still further to the credit of the Musselburgh men, and everyone thought that victory was theirs; but one never knows when the tide at golf will turn—and turn it did. Allan warmed up and got more into his game; and then one hole was taken and another and yet another; and I remember Captain Campbell of Schiehallion, with whom I was walking, saying in great glee—'Gad, sir, if they take another hole they'll win the match!' And, to be sure, another was won, and so on until the match stood all equal and two to play.

How different the attitude of the Dunns' supporters now from their jubilant and vaunting manner at lunch time! Silence reigned, concern was on every brow, the elasticity had completely gone from Gourlay's step, and the profoundest anxiety marked every line of his countenance. The very Dunns themselves were demoralised!

On the other hand, Allan and Tom were serene, and their supporters as lively as they had been depressed before. We felt victory was ours!

When the tee shots were played for the second last hole, off we flew as usual to see whose ball lay best! To our intense dismay Allan's lay very badly, whilst the Dunns' lay further on beautifully. Should the Dunns win this hole they would be DORMY—they might win the match! Our revulsion of feeling was great, and as play proceeded was intensified, for Allan and Tom had played three more with their ball lying in a bunker close to and in front of the putting green!

But, on the other hand, the Dunns' ball was lying close at the back of a curb-stone on a cart track off the green to the right! First of all they wished the stone removed, and called to some one to go for a spade; but Sir David Baird would not sanction its removal, because it was off the course and a fixture. The ball had therefore to be played as it lay. One of the Dunns (I forget which) struck at the ball with his iron but hit the top of the stone. The other did the same; and again the same operation was performed and 'the like' played. All this time the barometer of our expectation had been steadily rising and had now almost reached 'Sct Fair!'

The odd had now to be played, and this was done by striking

the ball with the back of the iron on to grass beyond the track. Had that been done at first, the hole might have been won and the match also; but both men had by this time lost all judgment and nerve, and played most recklessly. The consequence was the loss of the hole, and Allan and Tom DORMY.

We felt the victory was now secure: and so, in fact, it turned out, and Allan and Tom remained the victors by two holes.

I think it only just to say that, in my opinion, the winning of the above match was due to Tom Morris. Allan was decidedly off his game at the start, and played weakly and badly for a long time— almost justifying the jeers thrown at him, such as 'That wee body in the red jacket canna play gouf', and such like. Tom, on the other hand, played with pluck and determination throughout.

H. THOMAS PETER, Golfing Reminiscences of an Old Hand, 1890

TAYLOR v. KIRKALDY

J. H. Taylor's first major win had been over Andrew Kirkaldy in a home-and-away match over Winchester and Burnham. He then compounded the insult in 1894 by being the first Englishman to win the Open Championship. Now, in the following year, he had come north once again for the Open (which he won for the second time), but on the way he issued a challenge to play anyone for a stake of £50 at St Andrews, even though he had never seen the course before. It is easy to see therefore that a good deal of Scottish pride was at issue, as well as the prize-money, and the victor's relish in his success is plain.

We played thirty-six holes and Taylor never got a hole in front of me from start to finish, and I never was more than one up. It was neck and neck all the way. The strain was 'gey bad'. There was one hole in the last round where he had a chance. The ground was hard and keen and a little wind was blowing. Both our balls lay on the green, not more than a foot from the hole and perhaps only ten inches. I said 'A half?' to Taylor. 'No, Andrew,' he said. 'You play.' His ball was about an inch in front of mine. I said 'Lift your ball, then.' I played and missed. The ball struck the side of the hole and dribbled a yard past on the keen green. It was like putting on a window. Taylor must, at that moment, have felt pleased that he decided to play instead of halving the hole. But holes are never lost or won till the ball is out of sight.

It was absurd to think that Taylor would miss a ten-inch putt after I had been daft enough to do it, so I played back to the hole very carelessly, feeling sure that he would get in; in fact I played with one hand on the back of the putter—tempting Providence! Luckily for me I did not miss that careless putt. When Taylor came to play his ten-inch putt we stood as still as tombstones. I cannot tell you what I felt like when he missed and ran two feet past the hole, and then missed again coming back. Of course I was pleased, as I had a right to be, and did not say 'Hard luck!' like a hypocrite . . .

On leaving the sixteenth green we were all even. Both had a satisfactory drive over the Stationmaster's garden going to the seventeenth green—one of the longest holes. Taylor played a fine second shot, his ball resting at the foot of the green. My second was away to the left behind the deep pot bunker. Some of my friends said, 'Pitch it over the bunker, Andrew.' I said, 'I dare not pitch this. If I do I'll put the ball in the road. Taylor has a certain 5 and I must get a 5.'

Mr John Ball, Mr Harold Hilton and Mr John Low and many other leading amateurs of the day were standing by weighing up the position. I noticed a little hollow to the right of the bunker and saw that if I played the shot properly I could cannon against the side and curl in towards the hole. But there was the risk of taking the wrong line and going into the bunker.

'Chance yer luck,' said John Herd, the uncle of Sandy Herd, who was my caddie.

'That's what I like to hear,' said I. 'No bunker-fright for me.' I chanced my luck and it came off. The shot was about twenty yards. I ran it up and the ball came beautifully round the bias of the ground and lay within two inches of the pin. 'Hard luck,' said somebody, thinking how near I was to holing out. But I had nothing to complain of with the ball lying where it did. The very ground seemed to shake with the clapping of the crowd . . .

We went to the eighteenth tee—the thirty-sixth of the match— and I being dormy crowed over him saying, 'That's the door locked, Taylor; you canna beat me now.' 'True, but I can draw with you, Andrew,' he said. 'It's possible, Taylor,' I replied, 'but I have only to get a half to win and you have to get a win to halve. . . .'

We both had long drives and Taylor's second lay near the green. He played the odd after we measured and almost holed a beautiful mashie pitch—the sort of shot at which he has been the master since that day, twenty-five years ago. The ball looked into the hole. It was the gutta ball, of course, that plays no tricks. My third shot was past the hole about a yard. I was in no hurry or flurry, but just looked and sank the ball, beating the champion by a putt. Taylor shook hands as heartily as if he had won.

ANDREW KIRKALDY, *Fifty Years of Golf*, 1921

OUIMET *v.* THE BRITISH

Francis Ouimet's victory over Vardon and Ray in a play-off in the 1913 US Open at the Brookline Country Club was immensely important in popular-izing the game in the States. For the first time the Americans had a hero of their own to cheer.

The faithfulness of his young caddie, Eddie Lowery, adds an extra touch of warmth to the story. It is interesting to speculate on the likely reaction of today's top pros to being told to keep their eye on the ball by a 10-year-old.

There is little to be said about the Championship itself. After the first three rounds, I was tied with both Vardon and Ray at 225. Ray was finishing his final round as I walked to the first tee. I watched him hole out and learned he had made a 79. It was raining, but, even so, a 79 did not seem very low. It made his seventy-two-hole total 304. Playing the fifth hole, I was told that Vardon had tied Ray. The rumours in championship play, particularly if you happen to be in the running, come thick and fast. I was next told Barnes had the Championship in the hollow of his hand. Then word came to me that Barnes had blown up.

I was having my own troubles out in that rain and nothing would go right. Out in 43, all hope seemed gone. Then someone said, 'Tellier will win in a walk.' The tenth hole was a par three. Owing to the sodden condition of the putting green, a high pitch was dan-gerous, because the ball would become embedded in the soft turf. I elected to use a jigger, intending to hit a low shot to the green. I forgot to look at the ball and hit it about fifteen feet. I put my next on the green eight feet from the hole and then took three putts for an inglorious five. Then I learned that Tellier had got into trouble and had finished behind both Vardon and Ray, who were still leading.

After that wretched five, walking to the eleventh tee between a lane of spectators, I heard one man say, 'It's too bad, he has blown up.' I knew he meant me, and it made me angry. It put me in the proper frame of mind to carry on. There was still a chance, I thought. People lined the fairway as I drove. A par four was helpful on the eleventh. A hard five on the twelfth helped not at all, because here was a hole where one might be expected to save a stroke, although it was a difficult four. Standing on the thirteenth tee, I realized I must play those last six holes in two under par to tie. There were two holes on the course where I thought I might save a stroke: the thirteenth, the one I had to play next, which was a drive and a short pitch, and the sixteenth, a short hole. I selected these two holes for reasons. I had been quite successful on the thirteenth and had scored threes there regularly. I had not made a two all the week, and I had an idea I should get one at the sixteenth. It was just an idea.

My drive to the thirteenth was satisfactory. With a simple pitch to the green, I mis-hit the ball and barely escaped a trap. My ball lay off the green thirty feet from the hole I had selected as one upon which to beat par. Instead of having a reasonably short putt, I was stuck with a chip shot. In any event, I chipped my ball right into the hole for my three and was still in the hunt. A routine five came on the long fourteenth. I missed my second to the fifteenth badly, so badly I missed every trap. I pitched on and got my par four.

Then came the sixteenth, the hole I had been expecting to make in two. I not only did not get my two there, but actually had to hole a nine-footer for the three. One of the last two holes had to be made in three, the other in four. They were both testing holes. As I splashed along in the mud and rain, I had no further ideas. I just wanted an opportunity to putt for one of those threes. I got it on the seventeenth. A drive and second shot played with a jigger placed my ball on the green fifteen feet from the cup. It was now or never. As I looked the line of putt over, I thought of one thing, giving the ball a chance—that is, getting it to the hole. I struck that putt as firmly as any putt I ever hit, saw it take the roll, bang smack against the back of the hole, and fall in for the three.

Now to get the four. A drive split the fairway to the last hole and was out far enough so that a long iron could reach the green. Eddie Lowery, my ten-year-old caddie, handed me an iron and said, 'Keep your eye on the ball and hit it.' I did. I lifted my head just in time to see the ball sail toward the pin, saw it land and, as I thought, kick forward, and I can remember saying to Eddie, 'I have a putt to win this Championship.' I was certain I had seen my ball clear the embankment and hop forward. As a matter of fact, the ball struck the top of the bank and stopped instantly just off the cut surface of the putting green. A chip shot left me a four-foot putt which I popped in. I had ended that seventy-two-hole stretch in a tie with Vardon and Ray for the Championship I had been most reluctant to enter.

Friends hustled me into the locker room building and the excitement was tremendous. One individual came to me and asked this question, 'Were you bothered while putting on the seventeenth green?' 'Not a bit,' was my reply. 'Why?' He went on to say that the highway directly behind the green was littered with motorcars, so much so that it was impossible for machines to move in either direction. Just then a motor came along, and the driver, seeing his path blocked completely, kept up a constant tooting of his horn as I was preparing to putt. I never heard a single sound, so thoroughly was my mind centred on the business of holing the putt.

After taking a bath, I walked home and turned in early for a real night's rest. I slept from nine-thirty until eight the next morning, and after a light breakfast hustled over to the Country Club for my play-off with Vardon and Ray. I did not feel nervous or unduly excited. I slipped on my golf shoes, got hold of Eddie Lowery, and went out to the Polo Field to hit a few practice shots. There was nobody about. The shots I hit felt fine. Soon some people came along and watched me. After perhaps a half-hour's practice I was told that Vardon and Ray were on the first tee waiting for the match to begin.

Johnny McDermott took my arm and said, 'You are hitting the ball well; now go out and pay no attention whatsoever to Vardon or Ray. Play your own game.' It was excellent advice and I promised Johnny I would do my best.

On the way to the tee my good friend Frank Hoyt ('Stealthy Steve') asked me if I would not permit him to carry my clubs. I had played much golf with Steve and he was a master in the finer points of the game. I told him he must see Eddie Lowery. He made one or two offers of money, but they did not tempt Eddie in the least. It was interesting to see the reaction of Eddie as he definitely and positively refused to be bought off. Finally Hoyt appealed to me. I looked at the ten-year-old Eddie; his eyes filled, and I think he was fearful that I would turn him down. In any event, he seemed so sincere I did not have the heart to take the clubs away from him, and my final gesture was to tell Steve Eddie was going to caddie for me.

It was raining, and the three of us were ushered into the tent near the tee to draw lots for the honour. I drew the longest straw and had to drive first. As I walked over to the sand-box, and realized what I was up against and saw the crowd, I was terribly excited. If I could only get my tee shot away! Eddie stepped up as he handed me a driver and said, 'Be sure and keep your eye on the ball.' The opening salute was a drive well down the middle of the fairway and for good length. Vardon and Ray followed suit. Ray was the only one who was long enough to reach the green on his second, but he sliced a brassie to the right.

We all got on in threes and took fives on the hole. I was left with a four-foot putt for my five, and I worried not a little over it. I tapped it in, and then almost instantly any feeling of awe and excitement left me completely. I seemed to go into a coma. Eddie kept telling me to keep my eye on the ball. He cautioned me to take my time. He encouraged me in any number of different ways. My first mistake was on the fifth hole, where the slimy grip turned in my hand and my second shot went out of bounds. But Vardon and Ray both erred on the same hole, and I was safe for the time being. Ray had taken a five on the third to our fours, and that was the only difference in the scores up to that point.

Vardon made the sixth in three and went into the lead. Ray was now trailing Vardon by two strokes and me by one. The seventh hole at Brookline is a hard par at three. Vardon was to the right of the green with his iron and needed four. I failed to lay a long approach putt dead, and took four. Ray was the only one to get a three and he pulled up on even terms with me.

The eighth hole was sensational. This hole measures three hundred and eighty yards, and the view of the green is more or less restricted by a hill. You can see the flag, but no part of the green. We all had fine drives. A tremendous crowd had gathered round the green to see the balls come up. I played my second with a mashie straight for the pin. In a few seconds a mighty roar went up. As I handed the club to Eddie, he said, 'Your ball is stone dead.' I wanted to think it was, but I wished also to prepare myself in case it was not. Therefore I said to Eddie, 'It is not stone dead, but I believe I shall have a putt shot for my three.' You see I did not wish to be disappointed.

As we walked towards the green and came to the top of the hill, I saw a ball twelve inches from the hole. It was mine. Ray was forty feet away with a sidehill putt and he tapped his ball as delicately as possible. It took the necessary turns and rolled right into the hole. Vardon had a four, and I got my three, which put us all even at the end of eight holes.

The next high light was the short tenth. This green was so soggy that both Vardon and Ray, after pitching on, had to chip over the holes made by their balls as they bit into the short turf and hopped back. I was fairly close in one. My opponents failed to make their threes, and I stepped into the lead by a stroke.

I added another stroke on the twelfth, where I got my four to their fives. Vardon dropped a nice putt for a three on the thirteenth, one under par, which brought him within a stroke of me. The long fourteenth was important. Ray might reach the green in two, but it was beyond the range of Vardon and myself. Ray drove last, and I saw him hurl himself at his ball to get just a little added length. When he played his second from the fairway he put every bit of power into the shot, but his timing was poor, and he hit the ball far to the right into a grove of chestnut trees. He recovered beautifully, and the hole was made in five by all.

I was paying as little attention as possible to the strokes of the others, because I did not wish to be unduly influenced by anything they did. I was simply carrying out McDermott's instructions and playing my own game. I could not help but notice, however, that Ray was struggling somewhat. I noticed, too, that Vardon, who seemed to be a master in mashie work, pulled his pitch to the green, which was not his natural way of playing such a stroke. Vardon normally played his pitches with a slight fade from left to right.

Ray got into all sorts of trouble on the fifteenth and he seemed out of the running. I never gave it a thought as he holed out in six. I still clung to my one stroke lead over Vardon through the sixteenth. Ray was now five strokes behind. Vardon had the honour on the seventeenth tee. This hole is a semi-dog-leg, and by driving to the right you eliminate all risk. On the other hand, if the player chooses to risk a trap on the left and gets away with it, he has a short pitch to the green. Vardon drove to the left. I saw his ball start, and that is all. I drove to the right. Ray tried to cut the trees on the left and hit a prodigious wallop that cleared everything, but his ball was in the long grass.

As we walked toward our balls, I saw that Vardon had caught the trap and his ball was so close to the bank he had no chance at all of reaching the green. He could just play out to the fairway. I knocked a jigger shot to the green, my ball stopping fifteen feet above the hole. Ray and Vardon took fives. As I studied my putt, I decided to take no liberties with the skid surface and simply tried to lay the ball dead for a sure four. I putted carefully and watched my ball roll quietly toward the hole. It went in for a three. With one hole left, I was now in the lead by three strokes over Vardon and seven over Ray.

The eighteenth hole was a hard two-shotter. The rains had turned the race-track in front of the green into a bog, and my one thought was to get over the mud. All hit fine tee shots. I placed my second on the green. It did not enter my head that I was about to become the Open Champion until I stroked my first putt to within eight or nine inches of the hole. Then, as I stepped up to make that short putt, I became very nervous. A veil of something that seemed to have covered me dropped from round my head and shoulders. I was in full control of my faculties for the first time since the match started, but terribly excited. I dropped the putt. Nothing but the most intense concentration brought me victory.

I was fearful at the beginning that I should blow up, and I fought against this for all I was worth. The thought of winning never entered my head, and for that reason I was immune to emotions of any sort. My objective was to play eighteen holes as well as I could and let the score stand for good or bad. I accomplished a feat that seemed so far beyond anything I ever hoped to do that, while I got a real thrill out of it, I felt I had been mighty lucky. Had I harboured the desire to win that Championship or an Open title of any kind, I might have been tickled beyond words. In sport one has to have the ambition to do things, and that ambition in my case was to win that National Amateur Championship. Therefore, I honestly think I never got the 'kick' out of winning the Open title that I might have done if I had thought I could win it.

FRANCIS OUIMET, *A Game of Golf*, 1933

SARAZEN v. HAGEN

Walter Hagen was by common consent the king of match players, so any win over him was particularly sweet. This meeting took place in the final of the PGA of 1923.

There has never been a golfer who could outthink and out-maneuver a match-player opponent as Walter Hagen could. You couldn't rattle Hagen, whatever you did. Throw a string of birdies in his face, and he'd smile that disturbingly undisturbed smile of his, and then hurl some birds of his own back at you, when it counted . . . .

I was edgy before my match with Hagen because I viewed it as the one and only opportunity I had to redeem my reputation. Hagen also meant business. He never liked to be without a title to place beneath his name, and in 1923 he had been stripped of the British and had nothing to wear in its stead. For another thing, I had incurred his ill will by ridiculing his stunt of making easy shots in exhibition matches look like Greek drama. For instance, Walter would have a lie in the wooded rough, with a nice opening to the green between two trees. He knew the moment he saw how his ball was lying that he would play it between those two trees, and eventually he did, ten minutes later, after his caddie had excavated every rock in the area and Walter had explored the wilderness for openings he hadn't the slightest intention of using. The gallery would go wild when he finally played a run-of-the-mill recovery through the obvious opening. These phony dramatics irritated me. As Hagen dawdled before his recovery I would chirp up impatiently, 'That's a simple shot, Walter, I'll walk ahead and meet you at the green.' We had another score to settle—this was our rubber match. I had beaten Walter over seventy-two holes in 1922. He had evened the score in Florida the next winter. . . .

I played offensive golf on the first nine of the last eighteen—I was out in 35—and with nine to go, I stood 3 up. Walter got one of these holes back with a great niblick on the twenty-ninth, but I held my ground after that and reached the thirty-fourth tee two holes to the good. Walter won the thirty-fourth. On the thirty-fifth, a par 5 that was neither exceptionally difficult nor exceptionally easy, both our drives were adequate. I played first and socked a brassie into the trap to the right of the green. I was sauntering down the fairway, watching Hagen over my shoulder, when I saw his second shot hook sharply and crash out-of-bounds. 'That does it!' I said happily to myself. 'He'll never be able to halve the sure 5 I've got.' Hagen coolly accepted the penalty and dropped a ball over his shoulder. He decided to stay with his brassie. This time he drilled a long, low, unwavering screamer that ran all the way to the green, twenty feet from the pin. I elected to play conservatively from the trap, blasting out cautiously and leaving myself a thirty-footer. My approach putt slid three feet past the cup. And then Hagen—he had it in the clutch, all right—rolled his twenty-footer into the very center of the cup. He had got down in 2 from 250 yards off the green. I stepped up to my three-footer. I needed it for a half now. It looked a lot longer than three feet, doubling its length the way all crucial putts do for the man who has to make them. I stroked the ball, stroked it well, but it twisted off the rim of the cup and hung on the lip. I had permitted Hagen to win a hole that I had expected to win. More than that, I had allowed him to erase the two-hole advantage I had held on the thirty-fourth tee. We were all square as we came to the home hole, both of us dour and determined.

Hagen's drive rolled into a trap guarding the green. Once again, mine found the corridor between the traps and finished on the apron. Hagen's recovery left him fifteen feet short of the cup. My chip trickled five feet past. Hagen putted for his birdie. Not a good putt. Off to the right. The relief I felt when I saw his ball veer off the line vanished immediately. Hagen's ball lay me a full stymie. I had no chance to go for the cup on my five-footer. Walter had sneaked off with a half.

Extra holes. No blood on the thirty-seventh. I dropped a three-and-a-half-footer for my half in 4. On to the thirty-eighth. Walter, still up, took the safe route on his tee-shot, placing it just beyond the sharp break to the left on this tree-lined dog-leg par 4. This set him up for an easy pitch to the green and a probable birdie. I decided to take my chance here. Boldly, perhaps foolishly, I went all out in an attempt to carry the trees in the V of the dog-leg. I hit the ball with a little more hook than I wanted and then heard a sickening crash. The ball had struck either the roof of the cottage in the trees or the trees themselves. A bad ricochet and I would be out-of-bounds. The best I could hope for was a playable lie in the wooded rough. This was the opening Walter had been waiting for. He walked briskly down the fairway.

My caddie uncovered my ball, safely in bounds. I had been lucky in one respect. The ball had caromed past the thickest cluster of trees in the angle of the dog-leg, and I had a fair opening to the green. On the other hand, my ball lay heavily matted by the tall, spiky growths in the rough, almost hidden from view. I braced myself with the firmest stance I could manage under the circumstances, and flailed my niblick through the rough and into the ball; I felt the blade catch the ball solidly and saw it fly out of the rough and kick onto the green, run for the pin, slow down and die out just two feet away. The shoe was on the other foot now. I looked over to see what dent that recovery had made in Walter's armor. He was visibly shaken. I had never seen Hagen lose his poise before and I doubt if any man in the gallery had. When he finally played his wee pitch he floofed it, like a duffer, into the trap between him and the green. His fighting instinct surged back then. He made a brave effort to hole his shot from the trap. When he failed by inches, he had lost the PGA Championship.

The hour after that match was the first and only time I have seen Hagen depressed. Usually the only sign by which you could tell when disappointment or dismay lay concealed behind Walter's oriental mask was the speed with which he gobbled his drinks when he hit the locker room. That afternoon concealment was beyond Walter's power. He was disconsolate and he was angry. 'Uh-h,' he groaned in disgust when we met in the locker room, 'you're the luckiest golfer who ever lived. I've seen a lot of lucky shots in my time, but today—I give up.'

'Whaddya mean, lucky?' I asked pugnaciously.

'I could name a million shots. That one on the thirty-eighth.'

'Look, Walter. You were pretty darn lucky yourself,' I came back. 'Don't think you halved the thirty-sixth through any great playing on your part. Without that stymie you'd have been a dead duck right there.'

I left him sitting on a locker room bench and telling his crestfallen court to leave him alone for a while.

GENE SARAZEN, *30 Years of Championship Golf*, 1950

HAGEN v. JONES

Hagen again, this time from his point of view. The match with Bobby Jones had a particular piquancy, since it was seen as pitting the leading amateur of the day against the leading professional. By 1926 Jones had won two US Amateur titles and a US Open; Hagen had won three PGA Championships, two British Opens, and two US Opens.

Although, as an amateur, Jones was not paid for competing, he was at the time a real estate salesman for a company with land near the Sarasota club. Hagen is said to have received $11,800 from the gate receipts, out of which, with his usual style, he donated $5,000 to charity and gave Jones a pair of diamond cuff-links.

One bit of psychology I worked out with Harlow which to me was most important. I wanted to play the first 36 holes on Bobby's course at Sarasota. I told my idea to Harlow, who had suggested that they toss a coin to decide where the first rounds would be played.

'Bob, if the crowd at Sarasota win the toss, act a little disappointed and upset, but agree to play there. But, if you win,' I said, 'give in to play the first 36 on their course because they were so nice to agree to meet Hagen.'

Harlow returned to Sarasota and made the final arrangements. Harlow was 'lucky' enough to lose the toss and Bobby elected to play the first thirty-six holes of our challenge match on the Sarasota course on the date set, February 28th, 1926.

Winning that challenge match was equally important to both Bobby Jones and to me. And certainly we each went into it determined to give it all we had in skill and experience. Bobby, as an amateur, could not receive any of the gate receipts at the Sarasota course or of the $5,000 guarantee which my friend Benjamin Namm of Brooklyn had put up for the match at Pasadena. Tickets sold for $3.30 and, while only 750 people paid admission, several hundreds crashed the gate at Sarasota to give their idol Bobby Jones the ardent support he rated. Bob Harlow and I would have been willing to give the money to charity, but Bobby suggested I take it. I was more than willing to do so when Mr Namm said that all amounts over the guarantee at Pasadena he would donate to charity.

I had the honour, stepped up to the first tee, and hooked a long drive to the edge of the woods. Bobby had a nice straight drive down the centre aisle. And so, with my good friend George Morse as umpire-referee, the match began.

The number one hole is a par 4. On my second I had a difficult four-iron to the elevated horseshoe green and so played conservatively to the short edge. Bobby, perhaps seeing a chance for an early lead, went boldly for the flag with his second shot. His ball went over the green, down the slope into the rough. Being away, I played first and put my shot up within eighteen inches of the cup. This left Bob a very difficult recovery shot, down a sloping green. His shot went past the hole some twenty feet and he missed his putt coming back. I was 1 up.

I had decided early that I would play safely on my opponent's home course, since I was on the defensive and should take care to get pars and keep Bob from building up any kind of a big lead. Because my putting had been so sharp in my practise rounds I felt I could afford to play short and get my pars the safe way.

Two holes stand out in my memory besides that opening hole, which was important because it blocked Jones from gaining the lead and building up too much confidence. I had obtained a lead of 2 up over him in the first nine holes. We halved the tenth and the eleventh, then he got down in one putt to win the twelfth with a birdie to my par 4. This reduced my lead to one hole. That decisive block came on the thirteenth hole. Off that tee, a short hole, Jones hit a fine spoon shot about twenty feet from the hole. I played a number one-iron hole high to the right about ten feet from the flat. Jones missed and I holed a 2 to regain my 2-up lead immediately. I felt a bit easier now. When the morning round of eighteen holes ended I was 3 up.

The next block occurred on the sixth hole of the afternoon round. Through the first five my lead zigzagged from 3 to 2 and back on each alternate hole, leaving me 3 up at the sixth tee. This sixth is a drive and a pitch. Bobby hit a long ball off the tee, just escaping the bunkers on the left side of the fairway. A single pine tree stands just to the right of this fairway . . . and I drove to a spot which left my ball completely stymied by this tree. I could play a mashie-niblick with a slice or take a chance with a straight-faced iron with the hope the ball would run through the bunker which crossed the entire front of the green. Jones played a good second shot on to the green within possible holing distance. He undoubtedly considered he had an excellent chance of winning this hole. If you had asked me then, I would have had to agree.

I elected to play the mashie-niblick and try to slice it but keep the ball in the air all the way to the green. I half-topped the shot but hit it so hard and gave it such a spin that the ball ran and ran, hopped through the bunker, climbed the bank and rolled to a stop a few feet beyond the pin. I holed a 3 and Bobby needed two putts and went 4 down. I'm certain my winning the hole with a shot like that didn't put Bobby or his followers in any comfortable frame of mind. But he won the seventh and I was back to a three-hole lead.

By the twenty-seventh I was still holding that 3 up, with nine holes to play. For the first time I decided to gamble with my lead, owing to the fact that even if Bobby won as many as four holes I'd be only one down going back to my home course at Pasadena. I'd become acclimated to the day and the conditions so I felt that some bold moves on my part wouldn't be amiss right then. I followed this decision and it sure paid off, for I played the last nine holes in 32. I picked up five more holes, leaving me 8 up at the finish of the first thirty-six holes.

Despite the fact that I personally felt tremendously elated from that eight-hole lead over Bobby, it did not arouse spectator interest in the final thirty-six holes played at Pasadena. Such a big lead seemed too uneven to make for an exciting day for the gallery.

On March 7th, the Sunday following the Sarasota match, we started on my home course. I had a putting streak at Pasadena that surpasses any I've had before or since. At Bobby's home course I'd been sharp in the putting department, too, for I'd used only twenty-seven putts in the first round and twenty-six in the second. Yet Bobby, a fine putter himself, took thirty-one putts in the first and thirty in the second round. He had taken eight more putts than I and perhaps that accounted for his being 8 down. By the end of the Pasadena rounds he had a total of eleven putts more than I and I had won 12 and 11 to play.

WALTER HAGEN, The Walter Hagen Story, 1957

WETHERED v. COLLETT

Joyce Wethered's own account of one of her matches with Glenna Collett, the American champion, makes a good contrast with the overt competitiveness of the men. It also provides an amusing follow-up to the 'What train?' story, which became something of a trademark of Miss Wethered's.

Of all the great players I have known, Glenna presents the most detached of attitudes in playing a match. She intrudes her presence to the smallest degree upon her opponents. I would even say that she appears to withdraw herself almost entirely from everything except the game, and her shots alone remind one of the brilliant adversary one is up against. If she is finding her true form then there is little hope, except by a miracle, of surviving—at any rate in an eighteen-hole match. But there are also some vague days in between, when her interest and concentration seem to be elsewhere. Her charm, however, to my mind as a golfer and a companion lies in a freedom of spirit which does not make her feel that success is everything in the world. Those who are so generous in defeat are the people most to be envied.

At Troon the match anticipated between her and myself was worked up to such a pitch beforehand that, when the day came, one of two things was almost bound to happen. Either we should rise to the occasion or one of us would fail under the strain of it. As events happily turned out for me, I played the best golf that I have ever succeeded in producing. With the exception of two poorish putts I know that I have never played the rest of the game so accurately or so well before or since. I have never strung so many good shots together (or so faultlessly for me) even if I have been able to produce similar figures (that is, fours for the match) by other and less correct means. But because I was hitting the ball so surely I was able to avoid what might easily have occurred under the stress of the moment—the slipping of one or two important shots and perhaps the loss of the match as well.

I think this probably explains the reason why Glenna topped two drives at the sixth and ninth holes, strokes which as genuinely surprised herself, I remember, as they surprised everyone else. Unexpected as they were, they undoubtedly turned the tide in my favour, the second slip coming at a very crucial moment to make me one up after I had lost the lead at the short eighth by a weak putt. Up till that moment there had been nothing in it between us. Glenna had drawn first blood at the third, holing a very good putt for a three; I had drawn level at the fifth and taken the lead at the sixth only to lose it again at the short 'Postage Stamp.'

On turning for home one up I won the tenth owing to a loose third by Glenna, and the eleventh by holing a long putt. It was at this point that I again lived up to my curious reputation for not noticing trains. As was remarked in *The Times*: 'Miss Wethered holed a long curly putt for a three characteristically enough with an engine snorting on the line behind her.' But this time I was more fully aware of the reality of the train in question. It was puffing smoke in clouds behind the green in a way that could not very well be ignored. However, I was too well acquainted with the ways of a Scotch engine driver not to know that he was determined to wait to see the hole played to a finish before he continued with his goods to Ayr. Knowing this, there was little to be gained by my waiting. Besides, it was just possible that a train was not an unlucky portent. Whatever may be the truth of that supposition the putt made me three up and almost decided, I think, the result of the game. We halved the next two; at the fourteenth Glenna missed a short one and so gave me another hole; and a half at the fifteenth brought the game to a conclusion, four and three.   JOYCE WETHERED, *Golfing Memories and Methods*, 1934

HAGEN AND BYRON NELSON

Although the story must have gained something in the telling, this is both another instance of Hagen's cavalier approach to his fellow competitors and an illustration of the relaxed management of events in his day.

winning that first pro tournament is still, even today, one of the most difficult things for a young player, regardless of how talented he may be. Byron Nelson's baptism in the hazards of near victory makes one of golf's better horror stories. Unknown and unsure, thin, young, broke, nervous, married, armed only with a good, quick, upright swing, Nelson learned the hard way. He was playing along in the General Brock Open at Niagara Falls when, suddenly, at the end of the third round he was the leader.

On Sunday morning it took Nelson a while to adjust to the headlines in the papers. That was Byron Nelson they were talking about. Guy from Fort Worth. Then he looked closer at the stories, at the pairings for the last round. Good God. He was paired with Walter Hagen. He turned to his wife, Louise, and in something akin to a death rattle, he broke the news.

'I, uh, I have to play with, uh, Hagen today,' he said, as if to apologize for the 88 he was bound to shoot and the no-money they would leave with.

Now in a situation like that a pro got to the course early, and Nelson did. He chipped awhile, putted awhile and worried awhile.

He blushed a lot and hung his head and hid and kept checking to see if his trousers were buttoned. Finally it was time to tee off, but Walter Hagen, naturally, had not arrived yet. 'Looks like Mr Hagen is late again,' sighed the starter, unconcerned.

'Late?' Nelson said. Of course. Late was part of it then, what a real pro did to a rookie without penalty or disqualification; in fact, what Hagen usually did to everybody. Didn't he once send to the clubhouse for a folding chair so that Gene Sarazen, the man who introduced the sand iron and steel shafts, could sit down and rest while he, that cunning Hagen, studied a simple chip shot? Wasn't it Hagen who liked to psych guys out by strolling over to their bags, peeking in at their clubs, shaking his head mournfully and walking away? Others, like Horton Smith, just squinted peculiarly at the rookies until the sad young men worked themselves into incurable hooks. And still others, like Dutch Harrison, would sweet-talk a rookie out of his game. 'Man, can you massage that ball,' Dutch would say. 'I ain't seen a swing that good since Mac Smith.' But prince of the Slow Plays, that was Hagen.

And so it went for Nelson, the starter telling him, 'Go ahead and tee off with Willie Goggin, Byron, or with Ed Dudley. We'll pair Mr Hagen with someone else when he arrives.'

'But Walter Hagen's my idol. I've wanted to play golf with him all my life,' Byron said, torn between high privilege and dire necessity.

'Well, we're very sorry.'

'But I'm leading the tournament,' Nelson said.

'Yes, we know.'

Nelson practiced some more and paced and sat and paced and putted, his slacks ballooning in the Niagara breeze like the Graf Zeppelin, his wide-bottomed tie whipping past the curled-up collar tips of his dollar-nineteen shirt and on around his neck.

Fidget, pace, putt, stroll, throw up went Nelson until, thanks a million, two hours later along came Hagen in his white-on-white silk shirt, his gold cuff links and more oil on his hair than they were pumping out of the East Texas fields.

'Hi, boy,' he said.

'It's—it's a real big honor,' said Nelson in a trance, a trance he did not recover from until forty-two strokes later on the front nine holes.

The disaster was not total. Nelson somehow rallied himself,

shot a 35 on the back for a 77 and salvaged second place in the
tournament.   DAN JENKINS, *The Dogged Victims of Inexorable Fate*, 1970

SAM SNEAD

The other method of supplementing the pro's income was hustling. Sam Snead
was clearly a master of this art, and, as he said himself, wherever the pigeons
were landing, he was never far away.

Dutch Harrison, the Arkansas Traveler, eventually became a consistent money-winner, but he spent his first six years without earn-
ing a single penny of official cash. He lived off the fat of other
men's golfing egos. In this respect he was much like Leonard
Dodson, or better still, the brilliant trickshot artist, Joe Ezar, from
Waco, Texas, who could make a golf ball sit up and speak. Ezar
would stow away on freighters to Europe and hustle his way back
on the Queen Mary balancing golf balls on top of each other on
a bet. He could balance one ball on top of another—Lord knows
how—hit the bottom ball onto a green and catch the other in his
hand before it reached the ground. For money. Ezar would turn
up at a tournament armed only with a derby hat, an overcoat and
a pair of street shoes. 'Loan me the equipment, and I'll pay you
back double,' he would say, and do it.

Dutch was never greedy. All he and his pal, Bob Hamilton,
wanted was a couple of 'nine-dollar pigeons' a day, enough to make
expenses. Harrison had that splendid talent of being able to name
his score. If his opponent shot 71, Dutch shot 70. If his opponent
shot 80, Dutch squeezed out a 79 and convinced the guy how
unlucky he was.

One afternoon on the Coast in 1937, as Dutch and Hamilton
negotiated on the first tee for a game, a stranger asked if he could
play along. He had a raw swing and a country voice, like Dutch,
which meant he couldn't be all bad.

Hamilton, ever eager with loot in sight, said something subtle
like, 'How much you want to play for?'

'Well, I don't know much about betting,' the man said.

'You come along with us,' said Hamilton.

No more than a few holes had been played before Dutch
and Hamilton were distracted from their game with the nine-
dollar pigeons. For every good shot they hit, the stranger hit one
better.

'My, my, son, you sure got yourself a pretty swing there,' Dutch said. 'That old hook grip don't bother you none at all, does it?'

Birdie, birdie went the innocent fellow.

Presently, Harrison lashed a spoon shot into a green—a career shot—and he thought to himself, 'Now we see who the men are.'

To which the stranger put a two-iron inside of him.

'Bob,' Harrison said to Hamilton. 'We done got ourselves a-hold of something here.'

Later on, after Dutch and Hamilton had paid off, the young stranger said, 'Sure do thank you fellers. Say, what time you gonna be here tomorrow?'

'Son,' Harrison said, 'you work your side of the road, and we'll work ours.'

'That,' Dutch says today, 'is how I met Sam Snead.'   Ibid.

SNEAD *v.* GONZALES

No doubt it was this kind of early training that prepared Snead for his match with Gonzales.

'Hustling' suckers for side money didn't interest me, since tournament winnings kept coming my way. After my 1937 rookie year on tour, I was lucky enough to lead the P.G.A. prize list of 1938 with nearly $20,000 won. Going into the Miami Open of the following winter, I had a string of solid finishes behind me. But I didn't exactly duck when the Mississippi marbles were rolling on the rug or a golf-betting proposition came along. At the St Paul Open, my shot average of 69.2 per round was good for only third place; at Milwaukee a 67.9 average had brought me a fourth. Only a dozen or so of the seventy-five to one hundred circuit pros were breaking even or better. The talent was bunched so tight that I gave myself only five more years to stay in the money-running.

At Miami my 271 score was good for first place, and next morning I received a summons from L. B. Icely, president of the Wilson Sporting Goods Company. In the golf industry, Mr Icely was Mr Big.

At the hotel he looked worried. 'Let's go upstairs to Tommy Armour's room,' he said. 'We have an emergency.'

Armour, the former American and British Open champion and a top teacher, knew who I was, but since Hillbilly Snead never had won a major title, there was no telling what he really thought of my game.

Icely explained the situation:

In New York, the well-known sportsman, Tommy Shevlin, and a Cuban sugar king named Thornwald Sanchez had turned a few barroom drinks into an argument—Sanchez claiming that no man alive could beat Rufino Gonzales, the pro at the Havana Country Club, on his home course. Plenty of challengers had tried. Nobody had made the grade.

'Well, I can find a man who can do it,' said Shevlin.

'For money or talk?' said Sanchez.

'For $5,000,' said Shevlin, 'unless you want to cover more.' They shook hands on it.

A while later Shevlin arrived in Havana with his wife and Sanchez jumped him. 'Where's that unbeatable pro you were bringing?'

Shevlin said, 'By golly, I forgot all about the bet.'

'Well, you'd better get busy,' said Sanchez. 'I've spread that $5,000 you laid me amongst my friends and we're ready to collect. Gonzales is waiting and there's no backing out. You'll have to forfeit the $5,000 unless you get a man up, and fast.'

'Forfeit, hell!' roared Shevlin.

He'd burned up the phone cables to Icely, and now Icely was asking Armour which U.S. pro he should rush to his friend's rescue.

Icely had a list of five names: Craig Wood, Byron Nelson, Jimmy Demaret, Ben Hogan, and Sam Snead.

Armour was in a ticklish spot. He knew that the $5,000 probably now amounted to ten times that much, the way Shevlin and his New York crowd and Cuban golf plungers liked to chunk it in, and a wrong guess wouldn't be popular with a lot of prominent people.

'Sorry to bother you,' said Icely, 'since we're on opposite sides of the fence, Tommy. You're with the MacGregor company and I'm Wilson, but there's so much at stake that I need help with the selection.'

'What's wrong with that man right there?' said Armour, nodding at me.

'This is for high stakes,' Icely said. 'So think about it hard.'

'No thinking necessary,' said Armour, paying me quite a compliment. 'Just send this man right here.'

That was Monday. Early next morning, over I went, with my hands already sweating, since I'd never seen a Cuban course and now faced thirty-six holes of medal play with maybe $100,000 riding on me. In Havana the cockfights and jai alai stopped drawing; everybody was betting on the match. Fights broke out at the Havana Country Club and downtown between natives and U.S. tourists. One Cuban was kicked out of the Havana C.C. because he backed me. An 'act of treason,' the board of governors called it.

Shevlin was on pins and needles. 'How much of the bet do you want?' his crowd asked me.

'Betting Sanchez wasn't my idea,' I came back. 'And I've never laid eyes on this Gonzales. I hadn't figured to get in on it at all.'

'Well, you've got $250 of it,' they said. 'We'll feel better about it that way.'

I got their idea: by holding a piece of the action whether I wanted or not, I'd train and practice harder.

It was agreed that Rufino Gonzales and I would play some practice rounds together so that the gamblers could draw comparisons. He was a cool-looking gent, medium build, about 170 pounds, with a flat backswing and very straight off the tee. His putting was amazing. On the slick, sun-cooked greens, the Cuban could putt rings around me. The worst hazard of all was the light. Due to what they called a peculiar refraction effect of the sun, a newcomer had a tendency to see double and misjudge distances. No wonder Gonzales never lost at home.

All the Yankees began to pray for rain.

We got it, at the last minute. Under gray skies, the biggest gallery in Cuban history gathered around the first tee. There were some tough-looking people in the crowd—the kind you wouldn't want around your henhouse after dark.

'Who are they—smugglers?' somebody asked.

'No—Batista's boys,' somebody else said. 'The dictator, himself, has some pesos down on Gonzales.'

Things like that don't help your peace of mind. I could see myself pulling a ball into a palm forest and, if I was leading at the time, not coming out.

After three holes, Gonzales was trailing by 1 stroke, but on No. 4 my ball was all over the carpet for 3 putts. The Cubans cheered and clapped the misses. Rufino squared the match. On the next hole, which I birdied to Rufino's par, the Americans went crazy.

Both sides were glaring blue blazes at each other. I was what you'd call half popular all the time.

Gonzales was dropping long and short putts, never relaxing the pressure, and I remembered that in practice a way had shown itself to equalize him. On the par-5 holes and one long par 4, a power hitter could cut the corner over some trees and save a stroke. The Cuban was 50 yards shorter off the tee than me. On doglegs, he had to take the long way home.

Swinging as hard as I ever have put me on the par-5 greens in 2 shots to his 3 for birdies, and on the par-4 it was the same story. Once I got the hang of the fast greens, that did it. We finished the first eighteen holes with a 71 for Gonzales, a 69 for me.

'Sensational!' said Shevlin. 'Another $500 of the bet is yours, and this time it's on the house.'

Any time there's an extra money chance, I come out punching, and it didn't slow me down over the final eighteen. It was Rufino who lost his coolness. Luckily, his famous putting ability wasn't there that last round, and he was out in 38 to my 34 before pulling himself together for a fine 33 coming back. His second round of 71 wasn't good enough—I had a 68—and when the last ball dropped, the Cubans were a mighty miserable and angry group. All the way to the clubhouse, jostled by Batista's mugs, the hair stood up on my neck. Havana papers estimated that well over $100,000 was lost, all because Shevlin and Sanchez had gotten into a barroom argument.

'Good luck, pal,' I told Gonzales. He had to stay and face his backers, including Batista. I got out by the first plane.

Back in Miami, Mr Icely, Tommy Armour, and others gave me a whooping reception. 'You feel all right, Sam?' they asked.

'No, pooped,' I said. 'That took too much out of me. Next time the stakes are that high, send Demaret, will you? He likes travel and would make a prettier corpse than me.'

But I knew I'd see a lot more of betting and would be performing under the gun many times. One rule I made was never to bet for fun. That's a national habit of American golfers, but with me gambling always had been a dead-serious thing. Porky Oliver said that once I was told a funny story about betting and laughed until you could have heard a pin drop. Well, where's the joke in losing your hard-earned cash? With me, nothing hurts worse.

SAM SNEAD, *The Education of a Golfer*, 1962

[143]

SARAZEN'S OPEN

Gene Sarazen's account of his win at the Open at Prince's in 1932 is another tribute to the importance of the caddie as guide, counsellor, and friend, as well as bag-carrier to the player.

In 1932 the British Open was scheduled to be held on the links of the Prince's Club in Sandwich, right next door to the Royal St George course where I had lost and Hagen had won in 1928. Daniels would be available at Prince's if I decided to take another crack at the British Open, but I wasn't so certain I cared to knock my brains out any more trying to win that ornery championship. I had failed at Muirfield in 1929, and in 1931 at Carnoustie, where I had played pretty fair golf, I again trailed the winner, Tommy Armour, by two strokes. There were other considerations. Money was scarce in 1932 and getting scarcer, and I was in no mood to squander away the bank account I had slowly been able to build up through my labors on the winter circuit.

It was Mary who decided that I should go over for the British Open. One evening that spring, after I had returned from Lakeville— I had changed my club affiliation from Fresh Meadow to Lakeville in 1931—Mary sat me down in our living room and assumed her best I-am-talking-business-so-be-prepared-to-take-me-seriously voice. 'Gene, I don't believe that I've ever seen you playing better than you are right now,' she said, starting out on a very good foot. 'I know how hard you've been working on keeping in condition, running up and down the front-hall steps, swinging that heavy club, morning, noon, and night, cutting yourself down to one cigar a day. Now don't interrupt me, Gene. Last week I was talking with Tom and Frances Meighan and they agree with me that your golf is better than it ever was and that you ought to play in the British Open.'

'That's all very well and good,' I answered, 'but what about the financial side of such a trip? This isn't any time to throw away a couple of thousand, Mary. It would cost me just about that to cover all my expenses. I don't see it.'

'I've given that a lot of thought, too,' Mary said. 'I decided it would be a good investment, everything considered.' She stopped for a moment and felt me out with a smile. 'Now, Gene, I've got your tickets and your hotel reservations all taken care of. The only thing you have to do is get your passport fixed up. You're sailing a week from tomorrow on the *Bremen*.'

I had a smooth crossing, and an enjoyable one, thanks to the live-ly company of Fred Astaire. In London the first man I ran into at the Savoy was Roxy, a crony from Lakeville and a fanatic golf fan. Roxy was going to play at Stoke Poges the next day and per-suaded me to come along.

When we arrived at Stoke Poges, a young caddy—he was about twenty-seven, a stripling among British caddies—grabbed my bag. We whistled around the course in 67. 'I'm going to caddy for you in the Open,' the young man informed me when the round was over. 'I know just the type of caddy you need, Mr Sarazen.'

My mind flashed back to Daniels. 'You're a very smart caddy,' I told the aggressive young man. 'But I've already got a caddy for the Open. Skip Daniels.'

'Oh, I know Daniels. He must be around sixty-five now.'

'Just about,' I replied.

'He's too old to carry this bag,' the young caddy continued. 'His eyesight is gone. On top of that I've heard that he's been ill. Why don't you let me caddy for you at Prince's? I don't want to run Daniels down, but you'd ruin your chances if you took him on. The way you played today, you can't miss.'

He had something there. That 67 had been as solid a round as I had ever played in England. If I could keep that up, no one would touch me in the Open.

I told the young man to meet me at Prince's.

After a few days in London, I went down to Prince's to practice. The first person I met, right at the gate, was Daniels. He was over-joyed to see me. While we were exchanging news about each other, I could see that the last four years had taken a severe toll of him. He had become a very old man. His speech was slower. That shaggy mustache of his was much grayer, his limp was much more obvious. And his eyes, they didn't look good.

'Where's your bag, sir?' Daniels asked, hopping as spryly as he could toward the back seat of my auto.

'Dan,' I said—I couldn't put it off any longer though I almost didn't have the heart to say it, 'Dan, this bag is too heavy for you. I know you've been in bad health, and I wouldn't want you to try and go seventy-two holes with it.'

Dan straightened up. 'Righto, sir, if you feel that way about it.' There was great dignity in the way he spoke, but you couldn't miss the threads of emotion in his voice.

[145]

'I'm sorry, Dan,' I said, and walked away. I had dreaded the thought of having to turn old Dan down, but I had never imagined that the scene would leave me reproaching myself as the biggest heel in the world. I attempted to justify what I had done by reminding myself that business was business and I couldn't afford to let personal feelings interfere with my determination to win the British Open. It didn't help much.

I was a hot favorite to win. The American golf writers thought that I had a much better chance than Armour, the defending champion, and the veteran Mac Smith, the other name entry from the States. George Trevor of the *New York Sun*, for example, expressed the belief that 'Prince's course, a 7,000-yard colossus, will suit Sarazen to a tee, if you will pardon the pun. It flatters his strong points—powerful driving and long iron second shots.' The English experts were likewise strong for me until, during the week of practice, they saw my game decline and fall apart. The young caddy from Stoke Poges did not suit me at all. I was training for this championship like a prizefighter, swinging the heavy club, doing roadwork in the morning, practicing in weather that drove the other contenders indoors. My nerves were taut and I was in no mood to be condescended to by my caddy. He would never talk a shot over with me, just pull a club out of the bag as if he were above making a mistake. When I'd find myself ten yards short of the green after playing the club he had selected, he'd counter my criticism that he had underclubbed me by declaring dogmatically, 'I don't think you hit that shot well.' I began getting panicky as the tournament drew closer and my slump grew deeper. I stayed on the practice fairway until my hands hurt.

Something was also hurting inside. I saw Daniels in the galleries during the tune-up week. He had refused to caddy for any other golfer. He'd switch his eyes away from mine whenever our glances met, and shuffle off to watch Mac Smith or some other challenger. I continued, for my part, to play with increasing looseness and petulance. The qualifying round was only two days off when Lord Innis-Kerr came to my hotel room in the evening on a surprise visit. 'Sarazen, I have a message for you,' Innis-Kerr said, with a certain nervous formality. 'I was talking with Skip Daniels today. He's heartbroken, you know. It's clear to him, as it's clear to all your friends, that you're not getting along with your caddy. Daniels thinks he can straighten you out before the bell rings.'

I told his Lordship that I'd been thinking along the same lines myself. Daniels could very well be the solution.

'If it's all right with you, Sarazen,' Lord Innis-Kerr said as he walked to the door, 'I'll call Sam the caddymaster and instruct him to have Daniels meet you here at the hotel tomorrow morning. What time do you want him?'

'Have him here at seven o'clock. . . . And thanks, very much.'

Dan was on the steps of the hotel waiting for me the next morning. We shook hands and smiled at each other. 'I am so glad we're going to be together,' old Dan said. 'I've been watching you ever since you arrived and I know you've been having a difficult time with that boy.' We walked to the course, a mile away. Sam the caddymaster greeted me heartily and told me how pleased everybody was that I had taken Daniels back. 'We were really worried about him, Mr Sarazen,' Sam said. 'He's been mooning around for days. This morning he looks ten years younger.'

Dan and I went to work. It was miraculous how my game responded to his handling. On our first round I began to hit the ball again, just like that. I broke par as Dan nursed me through our afternoon round. We spent the hour before dinner practicing. 'My, but you've improved a lot since 1928!' Dan told me as he replaced my clubs in the bag. 'You're much straighter, sir. You're always on line now. And I noticed this afternoon that you're much more confident than you used to be recovering from bunkers. You have that shot conquered now.' After dinner I met Dan by the first tee and we went out for some putting practice.

The next day, the final day of preparation, we followed the same pattern of practice. I listened closely to Dan as he showed me how I should play certain holes. 'You see this hole, sir,' he said when we came to the 8th, 'it can be the most tragic hole on the course.' I could understand that. It was only 453 yards long, short as par 5's go, but the fairway sloped downhill out by the 200-yard mark, and eighty yards before the green, rising twenty-five to thirty-five feet high, straddling the fairway and hiding the green, loomed a massive chain of bunkers. 'But you won't have any trouble on this hole,' Dan resumed. 'You won't have to worry about the downhill lie on your second shot. You have shallow-face woods. You'll get the ball up quick with them. I should warn you, however, that those bunkers have been the graveyard of many great players. If we're playing against the wind and you can't carry them, you must play safe. You cannot recover onto the green from those bunkers.' Yes, I thought as Dan spoke, the 8th could be another Suez.

That evening when the gathering darkness forced us off the greens and we strolled back to my hotel, Dan and I held a final powwow. 'We can win this championship, you and I,' I said to Dan, 'if we do just one thing.'

'Oh, there's no doubt we can win it, sir.'

'I know, but there's one thing in particular we must concentrate on. Do you remember that 7 at the Suez Canal?' I asked.

'Do I!' Dan put his hand over his eyes. 'Why, it's haunted me.'

'In this tournament we've got to make sure that if we go over par on a hole, we go no more than one over par. If we can avoid taking another disastrous 7, Dan, I don't see how we can lose. You won't find me going against your advice this time. You'll be calling them and I'll be playing them.'

Mac Smith and Tommy Armour were sitting on the front porch when we arrived at the hotel. 'Hey, Skip,' Armour shouted. 'How's Eugene playing?'

'Mr Sarazen is right on the stick,' Dan answered, 'right on the stick.'

The qualifying field played one round on Royal St George's and one on Prince's. There isn't much to say about my play on the first day at Prince's. I had a 73, one under par. However, I shall never forget the morning of the second qualifying round. A terrific gale was blowing off the North Sea. As I was shaving, I looked out of the window at the Royal St George's links where I'd be playing that day. The wind was whipping the sand out of the bunkers and bending the flags. Then I saw this figure in black crouched over against the wind, pushing his way from green to green. It was Daniels. He was out diagramming the positions of the pins so that I would know exactly how to play my approaches. I qualified among the leaders. You have to play well when you're partnered with a champion.

The night before the Open, the odds on my winning, which had soared to 25-1 during my slump, dropped to 6-1, and Bernard Darwin, the critic I respected most, had dispatched the following lines to

The Times: 'I watched Sarazen play eight or nine holes and he was mightily impressive. To see him in the wind, and there was a good fresh wind blowing, is to realize how strong he is. He just tears that ball through the wind as if it did not exist.'

On the day the championship rounds began, the wind had died down to an agreeable breeze, and Daniels and I attacked from the very first hole. We were out in 35, one under par, with only one 5 on that nine. We played home in 35 against a par of 38, birdieing the 17th and the 18th. My 70 put me a shot in front of Percy Alliss, Mac Smith, and Charlie Whitcombe. On the second day, I tied the course record with a 69. I don't know how much Dan's old eyes could perceive at a distance, but he called the shots flawlessly by instinct. I went one stroke over on the 9th when I missed a curling 5-footer, but that was the only hole on which we took a 'buzzard.' We made the turn in 35, then came sprinting home par, par, birdie, par, par, birdie, birdie, birdie, par. My halfway total, 139, gave me a three-shot margin over the nearest man, Alliss, four over Whitcombe, and five over Compston, who had come back with a 70 after opening with a 74. Armour had played a 70 for 145, but Tommy's tee-shots were giving him a lot of trouble—he had been forced to switch to his brassie—and I didn't figure on too much trouble from him. Mac Smith had started his second round with a 7 and finished it in 76. That was too much ground for even a golfer of Mac's skill and tenacity to make up.

The last day now, and the last two rounds. I teed off in the morning at nine o'clock. Three orthodox pars. A grand drive on the 4th, and then my first moment of anguish: I hit my approach on the socket. Daniels did not give me a second to brood. 'I don't think we'll need that club again, sir,' he said matter-of-factly. I was forced to settle for a 5, one over par, but with Daniels holding me down, I made my pars easily on the 5th and the 6th and birdied the 7th.

Now for the 8th, 453 yards of trouble. So far I had handled it well, parring it on both my first and second rounds. Daniels had given me the go-ahead on both my blind second shots over the ridge of bunkers, and each time I had carried the hazard with my brassie. On this third round, I cracked my drive down the middle of the billowy fairway. Daniels handed me my spoon, after he had looked the shot over and tested the wind, and pointed out the direction to the pin hidden behind the bunkers. I hit just the shot we wanted—high over the ridge and onto the green, about thirty feet from the cup. I stroked the putt up to the hole, it caught a corner and dropped. My momentum from that eagle 3 carried me to a birdie 3 on the 9th. Out in 33. Okay. Now to stay in there. After a nice start home, I wobbled on the 411-yard 13th, pulling my long iron to the left of the green and taking a 5. I slipped over par again on the 335-yard 15th, three-putting from 14 feet when I went too boldly for my birdie putt and missed the short one coming back. I atoned for these lapses by birdieing the 16th and the 18th to complete that long second nine in 37, one under par, and the round in 70, four under. With eighteen more to go, the only man who had a chance to catch me was Arthur Havers. Havers, with 74–71–68, stood five strokes behind. Mac Smith, fighting back with a 71, was in third place, but eight shots away. Alliss had taken a 78 and was out of the hunt.

If the pressure and the pace of the tournament was telling on Dan, he didn't show it. I found him at the tee after lunch, raring to get back on the course and wrap up the championship. We got off to an auspicious start on that final round—par, birdie, par, par. On the 5th I went one over, shook it off with a par on the 6th, but when I missed my 4 on the 7th I began to worry about the possible errors I might make. This is the sure sign that a golfer is tiring. The 8th loomed ahead and I was wondering if that penalizing hole would catch up with me this time. I drove well, my ball finishing a few feet short of the spot from which I had played my spoon in the morning. Daniels took his time in weighing the situation, and then drew the spoon from the bag. I rode into the ball compactly and breathed a sigh of relief as I saw it get up quickly and clear the bunkers with yards to spare. 'That's how to play golf, sir,' Daniels said, winking an eye approvingly. 'That's the finest shot you've played on this hole.' He was correct, of course. We found out, after climbing up and over the ridge, that my ball lay only 8 feet from the cup. I holed the putt for my second eagle in a row on the hole, and turned in 35, after a standard par on the 9th.

Only nine more now and I had it. One over on the 10th. Nothing to fret about. Par. Par. Par. A birdie on the 14th. Almost home now. One over on the 15th, three putts. One over on the 16th, a fluffed chip. Daniels slowed me down on the 17th tee. 'We're going to win this championship, sir. I have no worries on that score. But let's make our pars on these last two holes. You always play them well.' A par on the 17th. On the 18th, a good drive into the wind, a brassie right onto the green, down in two for a birdie. 35–39–74, even par. There was no challenge to my total of 283. Mac Smith, the runner-up, was five shots higher, and Havers, who had needed a 76 on his last round, was a stroke behind Mac.

Feeling like a million pounds and a million dollars respectively, Daniels and I sat down on a bank near the first tee and congratulated each other on a job well done. Our score of 283–70, 69, 70, 74—was 13 under par on a truly championship course, and it clipped two strokes off the old record in the British Open, Bob Jones' 285 at St Andrews in 1927. (Incidentally, 283 has never been bettered in the British Open, though Cotton equaled that mark at Sandwich in 1934, Perry at Muirfield in 1935, and Locke at Sandwich in 1949.) Much as I was thrilled by setting a new record for a tournament that had been my nemesis for a decade, I was even more elated over the method by which I had finally reached my goal. I had led all the way. I had encountered no really rocky passages because I had had the excellent sense to listen to Daniels at every puzzling juncture. Through his brilliant selection of clubs and his understanding of my volatile temperament, I had been able to keep my resolution to go no more than one over par on any hole. The 8th, which I had feared might be a second Suez, had turned out to be my best friend. I had two 3's and two 5's on a hole on which I would not have been unwilling, before the tournament, to settle for four 6's. In fact, there wasn't one 6 on my scorecard for the four rounds of the championship. It is a card of which I am very proud.

After a shower, I changed into my brown gabardine jacket and was going over the acceptance speech I had prepared four years earlier, when the officials told me they were ready to begin their presentation ceremonies on the porch of the clubhouse. I asked them if it would be all right if Daniels came up and stood beside me as I received the trophy, since it had really been a team victory. They regretted to have to turn down a request they could sympathize with, but it was against tradition. I scanned the crowd gathering before the clubhouse, looking for Dan so that I could at least take him down front. I couldn't find him. Then, just as the officials were getting impatient about delaying the ceremony any longer, I spotted Dan coming down the drive on his bicycle, carrying a grandson on each handlebar. On with the show.

After the ceremony the team of Daniels and Sarazen got together for a rather tearful good-bye. I gave Dan my polo coat, and told
him I'd be looking for him the next year at St Andrews. I waved
to him as he pedaled happily down the drive, the coat flapping in
the breeze, and there was a good-sized lump in my throat as I
thought of how the old fellow had never flagged for a moment
during the arduous grind of the tournament and how, pushing himself all the way, he had made good his vow to win a championship
for me before he died.

It was the last time I saw Dan. A few months later some English
friends, who kept me posted on Dan, wrote me that he had passed
away after a short illness. They said that after the Open he had worn
the polo coat continually, even inside the pubs, as he told the golf
fans of three generations the story of how 'Sarazen and I did it at
Prince's.' When old Dan died the world was the poorer by one
champion. GENE SARAZEN, *30 Years of Championship Golf*, 1950

WATSON *v.* NICKLAUS

*Nobody who saw the extraordinary last two rounds of the Open at Turnberry
in 1977 will forget the magnificent intensity of the contest between Watson
and Nicklaus. If ever a tournament turned into a match, this was it, and
the quality of the golf lived wonderfully up to the occasion.*

Now to Watson and Nicklaus. No sooner had they teed off on
Friday, the day of the third round, than the air began to bristle.
Just about everyone in the immense gallery was thinking about
their classic encounter on the final nine holes of this year's Masters,
when Watson, playing right behind Nicklaus and completely aware
of what he had to do to stay with this tremendous finisher, refused
to be awed, and finally beat him by birdieing the seventy-first hole.
Nicklaus, who is now thirty-seven, has been the best golfer in the
world for the last twelve years, or possibly even longer, and he sees
no reason to cede his throne to any young pup, since he believes
that in many ways he is a better golfer today than he has ever been.
Watson, who is ten years younger than Nicklaus, is patently an
improved player this year. His left side controls his swing just the
way it should, and his footwork is very good—not that these are
the things you are most conscious of while watching him. What
impresses you most is the quickness and decisiveness with which
he plays his shots, the freedom of his hitting action, and the

sharpness with which he strikes the ball. Much as Watson respects Nicklaus, you felt at Turnberry that he relished the challenge of facing him head-to-head. On both days, he played the better golf from tee to green, but Nicklaus came through with so many stunning recoveries and got down so many long and difficult putts in key situations that until the very end he was the man in command and Watson the hard-pressed pursuer.

Perhaps the most effective way to get across the furious attacks and counterattacks that made this protracted confrontation between Watson and Nicklaus so memorable is simply to set down, with a minimum of embellishment, what took place on the significant holes. On Friday, Nicklaus, off fast, birdied the first hole, a 355-yard par 4, when he played his sand-wedge approach three feet from the cup. Watson answered this with a birdie 3 on the third, a 462-yard par 4; his second shot, a 5-iron, left him only a six-foot putt. On Woe-be-Tide, the par-3 fourth hole, both men hit 6-irons, Watson stopping his shot two feet from the pin. Nicklaus, however, holed from twenty feet. Two birdies. On the sixth, a 222-yard par 3 across a deep swale, Nicklaus moved out in front by two strokes. He smashed a 2-iron to twelve feet and made the putt. Watson, who was bunkered off the tee, missed his try for a par from eight feet. On the seventh, a 528-yard par 5 that moves uphill most of the way, Watson reached the green in two. Nicklaus, who was starting to pull his tee shots, was never on the fairway on this hole but curled in another good-sized putt—this one from twelve feet—to match Watson's birdie. Scores at the turn: Nicklaus, 31; Watson, 33.

Starting home, Nicklaus continued to pour it on, birdieing the tenth, a 452-yard par 4, with a twenty-five-foot putt. Watson then holed from ten feet for *his* birdie. The fourteenth is a 440-yard par 4 to an elusive green. Here Nicklaus made his first real error of the day, misreading the three-footer he had for his par. Watson, who had looked as if he might fall a stroke farther behind when his approach bounded over the green, got down in two from the rough with a perfectly gauged chip and actually picked up a stroke. Only one behind now. He got that stroke back on the next hole, the 209-yard fifteenth, when he knocked in a twenty-footer for a birdie 2. On the seventeenth, an eminently birdieable 500-yard par 5, both made their 4s. Nicklaus let a big chance slip away here: his second, a great 2-iron, finished three feet from the pin, but he pushed his putt for an eagle a shade off line. Scores for the round:

Nicklaus, 65 (31–34); Watson, 65 (33–32). Each had made six birdies and one bogey. Nicklaus was never behind on this round, but the two short putts he failed to make on the in-nine had cost him the lead.

On Saturday, Nicklaus was again off fast. He jumped into a two-stroke lead with a birdie on the second, a 428-yard par 4, by playing a remarkable 7-iron from an awkward sidehill stance in the rough (the ball was well above his feet) ten feet from the pin. Watson bogeyed the hole, pulling his approach to the left of the green and taking three to get down. Nicklaus increased his lead to three strokes with another 2 on Woe-be-Tide, holing this time from thirty feet. Watson then mounted a terrific rally. He cut away one stroke of Nicklaus's margin with a birdie 3 on the 411-yard fifth when he followed a lovely 5-iron with a sixteen-foot putt. He cut away another stroke on the long seventh when he reached the green in two, using a driver off the fairway, and two-putted for his birdie. On the eighth, a 427-yard par 4, Watson cut away the last stroke of Nicklaus's lead with yet another birdie: after a 5-iron approach to the heart of the green, he rapped in a twenty-footer. (I don't believe I have ever before seen two golfers hole so many long putts—and on fast, breaking, glossy greens. They were also doing such extraordinary things from tee to green that it was hard to believe what you were seeing.) Then, having come all the way back, Watson once again fell behind. On the ninth, the 455-yard par 4, where the tee is set on the isolated crag, he pushed his drive into the rough on the right, pulled a 1-iron across the fairway into the rough on the left, and needed three more shots to get down. Nicklaus saved his par when he eased a touchy downhill twelve-footer into the cup. Scores on the front nine: Nicklaus, 33; Watson, 34.

On the twelfth, a par 4 only 391 yards long, Nicklaus, after pitching with a wedge, rolled in still another long putt—a twenty-two-footer—for a 3. This birdie moved him out in front by two strokes, and many of us felt at that moment that we had perhaps watched the decisive blow of the struggle. After all, holes were running out fast now—only six were left—and how often does a golfer of Nicklaus's stature fail to hold on to a two-stroke lead as he drives down the stretch? Watson was not thinking in those terms. At this point, he rallied once again. He birdied the thirteenth, a shortish par 4, with a well-judged wedge shot twelve feet from the pin and a firm putt. Then he again birdied the fifteenth, the 209-yard par 3: using a putter from the left fringe of the green, he stroked a sixty-footer smack into the middle of the cup. Now Watson and Nicklaus were again tied. Everything to play for.

Two pars on the sixteenth. On the seventeenth, the 500-yard 5, Watson was down the fairway with his tee shot, a few yards shorter than Nicklaus, who was in the rough on the right. Considering the circumstances, Watson then played one of the finest shots of the day—a 3-iron that was dead on the pin all the way and finished twenty-five feet beyond it. Nicklaus went with a 4-iron. He hit it a trifle heavy, and the ball came down in the rough short of the green and to the right. Throughout these last two rounds, Nicklaus, on his frequent visits to the rough, had improvised a succession of astonishing shots to scramble out his pars and an occasional birdie, and here he punched a little running chip that climbed up and over two ridges in the green and came to rest three and a half feet past the cup. Watson was down in two safe putts for his birdie 4. Nicklaus, who hadn't missed a putt all day, missed his short one. The ball started left and stayed left and didn't touch the cup. A 5, not a 4, and at the worst time possible. Now, trailing Watson by a stroke, Nicklaus's only hope was to birdie the eighteenth, a 431-yard par 4 on which the fairway bends to the left. After Watson had played a prudent 1-iron down the middle of the fairway, Nicklaus, going with his driver, belted a long tee shot that kept sliding to the right— far to the right, and deep into the rough. He was faced with an almost unplayable lie: the ball had ended up two inches from the base of a gorse bush; and, to complicate matters, a branch of the bush, some two feet above the ground, was directly in the line of his backswing. He would be hard put to it to manufacture any kind of useful shot. Watson to play first, going with a 7-iron. He couldn't have hit it much better: the ball sat down two feet from the pin. Nicklaus now, playing an 8-iron. Still not giving up. With his great strength, he drove his club through the impeding gorse branch, managed to catch the ball squarely, and boosted it over the bunker on the right and onto the green thirty-five feet from the cup. He worked on the putt, studying the subtleties carefully, as if the championship depended on it. Then he rammed it in for his 3. Tremendous cheers. Watson chose not to give himself too much time to think about the two-footer he now had to hole to win. After checking the line, he stepped up to the ball and knocked it in. Another salvo of cheers—for Watson, and also for both Watson and Nicklaus and the inspired golf they had played over the last thirty-six holes of the Open. The scores for the concluding round: Watson, 65 (34–31); Nicklaus, 66 (33–33). (Nicklaus, incidentally, had only one bogey over the last two rounds.) Their scores for the tournament: Watson 68–70–65–65—268. Nicklaus, 68–70–65–66—269. Both shattered to smithereens the old record total for the British Open of 276.

For some reason or other, Jack Nicklaus always moves one more in defeat than in victory. I don't know exactly why this is, for he is an excellent winner. Anyway, he is probably the best loser in the game. He was direct and honest in his assessment of why Watson had won: Tom, he said, had played better golf than he had on both rounds. He had no alibis. Nevertheless, for all his self-possession, it was observable that Nicklaus had been hit hard by losing a second major championship to Watson this year on the final holes after giving the pursuit of winning everything he had. That last phrase is important. No matter what the odds are, Nicklaus never stops fighting, and you never know when he will contrive some small miracle like that impossible 3 on the last hole. In a word, I should say that over the past dozen years, without any question, he had been far and away the best competitor in any sport in which it is one individual against other individuals, not team against team.

How was Watson able to play so superlatively, to keep coming back so valiantly time after time? Alfie Fyles, his gnarled little caddie from Southport, near Liverpool, who also caddied for Watson when he won the British Open at Carnoustie in 1975, put it this way: 'His swing is better, more compact. His approach to the game has improved a great deal. He thinks better. He's a mature golfer now.' What underlies the confidence that Watson exhibited throughout his extended duel with Nicklaus? For one thing, his triumph in the Masters undoubtedly convinced him that he had reached the point where he could stand up to a giant like Nicklaus. For another, he is able to get himself up very high when he faces Nicklaus, for he appreciates that this is essential. He also knows that he is a sound driver and a sound putter, and, though he doesn't make this kind of pronouncement, he believes he hits the ball as well as anyone. Despite a sensitive temperament and an active imagination, he is now able to relax under pressure. This, of course, helps him to execute his swing well, and when a golfer like Watson feels that he is swinging just the way he wants to, it affirms his feeling of confidence and imbues him with an extraordinary resilience.

The general feeling at Turnberry after the Open was that the R. and A. is almost certain to award the Ailsa course a regular place on its championship rota. After all, if the course produced such a thrilling Open when it was in relatively poor shape, it is only logical to think that it will provide a first-rate test when it is its real self—bearded with some good Scottish rough, and swept by good Scottish winds that will bring the bunkers into play and make ball control critical. (It is, I admit, something of a mystery how the almost defenseless course yielded so few low scores—other than those by Nicklaus and Watson—on the last two and a half days. Some players I talked with thought that the hard-surfaced, rolling greens had effected this, making it difficult to stop one's approaches close to the pins and also to putt consistently well. This suggests that the two leaders, locked deep in their duel, simply forced each other to rise above the conditions.) Everything being equal, the British Open should be back at Turnberry in about seven years. By that time, we should have a much better idea whether Tom Watson is just a very talented golfer or a great golfer. While he is different from Bobby Jones in many ways, there is in him more than a touch of Jones. In seven years, Jack Nicklaus will be forty-four, and yet I wouldn't be at all surprised if at that distant date this prodigious athlete still loomed as a serious contender for the championship.   HERBERT WARREN WIND, *Following Through*, 1985

WOOSNAM WINS THE MASTERS

The US Masters at Augusta has from its very beginning developed a reputation for thrilling finishes. The brilliant layout of the course, which allows the player to choose between the safe shot and the heroic gamble on so many of its holes, and especially over the crucial last nine, almost guarantees the drama that has constantly been a part of this tournament. Sarazen's famous second shot with a four-wood for a two at the par-5 fifteenth, which enabled him to tie with Craig Wood and then beat him in the play-off in 1935; the battle between Watson and Nicklaus in 1977; Ballesteros winning and then later losing it in the water of Rae's Creek when apparently well clear of the field; Lyle's shot out of the bunker at the eighteenth in 1988; Faldo's successive play-off wins in 1989 and 1990—all of these have confirmed the Masters

[157]

*as the most reliably exciting of all the Majors. But perhaps none of these
great events is more remarkable than the manner of Woosnam's win in 1991.*

Up ahead Jose-Maria Olazabal was in trouble. He had gone from
the fairway bunker on the 72nd hole of Augusta National to the
bunker by the green.

Back on the tee, Ian Woosnam was in torment. He knew that
the Spaniard's erratic progress meant that a par would probably
give him his first major championship, the US Masters. The problem was how to get it.

He asked his caddie Phil Mobley how far it was to the fairway
bunkers. The question came out as a croak. Throat and tongue
dried by the tension, Mobley told him the distance.

Then Woosnam had his moment of inspiration. 'How far is it,'
he asked, 'to *carry* those bunkers?' Mobley, good caddie that he is,
quickly worked it out. 'It's 268 yards,' he said, 'to get over the lot.'

In those moments Woosnam and Mobley had initiated a new
strategy for playing the 18th at Augusta. There was nothing subtle about it: just blast the ball as hard as possible left, going left.

'I know,' says Woosnam, 'that when I hit it as hard as I can the
ball will either go straight or draw. It's natural to my swing. So
when I saw Ollie [playing up ahead] in the bunker and then my
playing partner Tom Watson went into the trees on the right, I
decided to go as far left as I could. There's nothing over there
except the members' practice area [which is in bounds but was
crammed with spectators] and the only possible problem would be
if a spectator walked off with my ball.'

Woosnam, one of the game's longest hitters, launched himself—
and hit it perfectly. As it rocketed off the tee the American television commentators immediately went into their 'Uh, oh' routine,
thinking it was headed for the sand. They were astonished when
it cleared everything, and so were the marshalls, who had not anticipated this new tactic.

The ball hurtled into a crowd of, literally, thousands, coming to
rest on trampled grass. Woosnam had negotiated the first, and the
worst, of the obstacles of the 18th hole.

Looking back at that act, after a year to think about it, Olazabal
is almost in awe of Woosnam's tee-shot. 'I had never even thought
of that route. I could not do that. Once you are there, of course,
it is perfect, but . . .'

It is difficult even now to persuade the Spaniard to talk about that last hole. So much happened to him that he feels was unfair. 'I hit a good solid tee-shot to the left centre of the fairway; it never moved in the air but when it landed it kicked left into the bunker. Then it was a seven-iron distance, but I was up against the face and had to take an eight; I hit it good, it pitched over the bunker by the green, and then it spins back into the sand. It was a tough bunker-shot but I hit it to 12 feet, before it came back and back, 40 feet away . . .'

Three good, solid shots, three poor results. He took a bogey five and knew as he walked from the green that his first chance of a major championship had almost certainly been taken from him.

Back down the fairway Woosnam was surrounded by bewildered spectators and frantic marshalls. In order for the Welshman to play his second they would have to clear away several thousand people and, as Woosnam says, 'they had no idea what they were doing. I had to join in, shout at people, which was the last thing I wanted in that situation.'

Even after five minutes of jostling he had only succeeded in clearing enough space to swing a club, with a narrow sightline through the crowds. 'I couldn't see the green; I couldn't even see the stands. I had to aim at a particular tree on the treeline 100 yards past the green.'

But all he had in his hands was an eight-iron; the lie was reasonable and he told himself to remember only the hours of practice. 'I knew I could do it with my eyes shut, which was just as well since it was a blind shot anyway.'

The shot was almost perfect. With another yard in length it would have been, but it finished just off the putting surface, leaving him with what in any other circumstances would have been a simple chip. He got it six feet past the hole and then, with that huge right uppercut, urged his ball into the hole.

Olazabal had watched the proceedings from the scorers' tent at the back of the green. When Woosnam's putt went in he was devastated. 'For me it was the first time I had felt I could win a major. One par at the last and I would be in the play-off, and I would not be afraid of anyone in a play-off.'

How upset was he? 'I cannot answer that. There are no words. But for months after that I was trying to make up to myself the disappointment. I was trying to hit perfect shots all the time, to try to win, to fill that hole.

'It was worse that it had happened at Augusta, the No. 1 course for me in the world. The skill that every player has can be seen at Augusta; it demands all the shots; nothing is taken away from you. If you have skill and imagination you can show it at Augusta.'

DAVID DAVIES, *Guardian*, 8 April 1992

GREAT COURSES

❖

*C*urrent *thinking divides courses into three categories—the penal, where the player is given no option but to attempt the most difficult shots, with dire penalties if he fails; the strategic, where the challenge is to manœuvre the ball around so as to make the next shot as easy as possible; and the heroic, where the brave shot is rewarded but an alternative, safer route is also available for the weaker or less intrepid player.*

All three of these categories are represented in this section: Oakmont and Pine Valley are clearly penal; Pebble Beach and Portrush somewhere between that and the traditional strategic links of St Andrews and Prestwick; and Augusta, as Bobby Jones makes plain, was planned from the beginning as heroic. Above all, reading the descriptions makes one long to play any and all of them again.

THE DIFFERENCE BETWEEN ENGLISH AND AMERICAN COURSES

The difference between the golf courses of America and of Great Britain can best be expressed by the two words 'artificial' and 'natural'; and that means a whole lot more than the mere presence or absence of the fabrication of man. The difference begins at the front of the tee and extends to the very back edges of the green, and it is a difference which a player cannot conquer by mechanical proficiency alone. To combat it requires judgment, experience, and, as O. B. Keeler would express it, 'something between the ears.'

To begin with, let us play a round upon the average first-class American course. American architecture allows practically no option as to where the drive shall go. To the expert, there is always the evident desirability of placing the ball on one side or the other of the fairway, but, as a rule, he is much more concerned with avoiding bunkers and long grass along the sides. There is only one safe place for the ball, and it must be put there. The shot is prescribed by the design of the hole—it must be in exactly a certain direction, and, usually, as long as possible.

Suppose then, that we have made a good drive and our ball is lying nicely in the fairway. From this situation, one of two shots is required—either a long wallop with a brassie with the hope of getting close to the green, or an iron or pitch shot designed to reach the green itself. The brassie shot, being simply an occasion for the application of a strong back, may be dismissed.

The iron shot, like the drive, leaves the player little choice. The green is usually well defined and sloped up toward the rear. Further, it is almost always well-watered, so that any shot with fair height will not go over if it pitches onto the green. The problem is to hit the green, and if that is done, it matters little what spin the ball may have. The soggy green takes the place of the perfect stroke.

Now, let me ask what manner of golfer will be developed by courses of this nature? The answer is—a mechanical shot producer with little initiative and less judgment, and ability only to play the shot as prescribed.

STRATEGY REQUIRED ON ENGLISH LINKS

Take this man and put him down for the first time on a British seaside links, and watch the result! There the fairways are wide, and the greens are watered only from the skies. The greens are quite unartificial, usually flat or even sloping away from the shot. Bunkers do not surround the green defining the target. In a word, there is no prescription. It is left entirely to the judgment and conscience of the player as to what route he will choose. If the greens are soft, the flag still offers a target, and our shot-making machine will not fare badly. But let there be dry weather for the real show!

I shall never forget how I cursed Hoylake in 1921, and St Andrews. At that time I regarded as an unpardonable crime, failure to keep the greens sodden, and considered a blind hole an outrage. I wanted everything just right to pitch my iron shots up to the hole. It was the only game I knew, so I blamed the course because I could not play it.

Later, I came to like this seaside golf just as any lover of the game must come to like anything which adds zest to the play. Variable conditions afford ample opportunity for the display of any strategic talent we may possess, and preserve in the most human of games that fascinating personal element which is its chief attraction.

ENGLISH COURSES NEED THOUGHT

On an American course we play—indeed, we are forced to play—
one stroke at a time. If we played the same hole twice a day for
months, we should always be striving to play it in the same manner, usually straight away from tee to cup. We should never change
tactics because of wind, rain or dry ground. We should rarely have
a choice, or an opportunity to think. Now, British seaside golf cannot be played without thinking. There is always some little favor
of wind or terrain waiting for the man who has judgment enough
to use it, and there is a little feeling of triumph, a thrill that comes
with the knowledge of having done a thing well when a puzzling
hole has been conquered by something more than mechanical skill.

And let me say again that our American courses do not require,
or foster, that type of golfing skill. Our men have to learn it in
England or Scotland, and that is the big reason they cannot play
there without experience.   BOBBY JONES, *Golf*

❖

ST ANDREWS

> Sweet saint whose spirit haunts the course
>   And broods o'er every hole,
> And gives the Driver vital force
>   And calms the Putter's soul. . . .

There is nothing new to say about St Andrews, just as there is
nothing new to say about Shakespeare; but wherever in the world
there is a natural masterpiece, critical man will try and 'murder to
dissect'. The Old Course is dissected critically or uncritically with
a vengeance! Fifty square feet—the size of a good big Board Room
table—*fifty square feet* of divots are taken, roughly, every day. Not
all are, alas, replaced. . . .

I must confess that I approached St Andrews sceptically, and
rather in the mood of the little girl in Landor's poem, seeing the
sea for the first time:

> all, around the child, await
> Some exclamation of amazement wild.
> She coldly said, her long lashed eyes abased,
> Is this the mighty ocean? Is this all!

I hope she was converted. I arrived at St Andrews by car from the
south—a good way to arrive, for, coming over the hill, the University

[163]

tower and the ruins of the Cathedral rise above a pastoral skyline before one sees the town—as do the heavenly towers of Chartres—and one's first glimpse is just these towers and beyond them the long stretch of white sand and sandhills, the sea, and in the far distance the north side of the Firth of Tay. One can imagine the Cathedral—it must have been superb in its prime—dominant on the rocky point, with the University under its wing, looking out north across the bay and north-west along that magnificent sand-beach to the point where the River Eden meets the sea. Geologists will tell you that at some remote time the sea receded from here, and that interaction of river and tide threw up sandbanks first; then, in time, 'firm soil won of the watery main'; seeds carried on birds' feet; seeds wind-blown; vegetation knitting the sand together, coarse marram grass and rush at the sea's edge, bent and fescue within-sides. Whins—gorse to the southerner—came, and heather; and by the fourteenth century stretching north from the town was a long low tract of seaside heath, full of rabbits and foxes and ideal for hunting and archery practice.

It was only forty-odd years after the Battle of Agincourt that golf appears officially in history, in an Act of the Scottish Parliament, 1457. St Andrews itself appears on January 25th, 1552. Archbishop Hamilton made a deed with the townspeople whereby the Church should farm the 'cunningis' (rabbits) and the townspeople have the right 'inter alia to play at golff, futball, schuting, at al gamis with all uther maner of pastyme as ever thai pleis'. It is curiously apt that bad golfers should be called rabbits.

True that the Honourable Company of Edinburgh Golfers did, in fact, formulate the first rules of golf which St Andrews copied ten years later, but the fact is that the Honourable Company was a small and select body amusing itself in a thriving capital city with a thousand other interests, while St Andrews is a small homogeneous town, and once it had taken to golf it loved golf with single-hearted passion; and also it possessed, in its links, a supreme golfing terrain, which Leith was not. It was utterly natural for St Andrews to become the genius of golf, and inevitable: life for its inhabitants thereafter was simple.

> The tee, the start of youth, the game our life,
> The ball when fairly bunkered, man and wife.

The first thing to do is to go out on to the links.

Is this the mighty Old Course? Is this all!

No, from the very start one is aware of greatness and strangeness. Everyone has seen pictures of the first tee in front of the Royal and Ancient club house, but they give only a faint flavour of the scene as it is played ordinarily. Golf photographs are mostly like first-night audiences, they do not give an impression of the play. But golf has been running at St Andrews for generations. There are the players ready to drive down the broad, wide, gently sloping expanse. Pedestrians begin to quicken across, accelerating like hens towards grain, either inland towards the houses or seaward towards the white rails that flank the right-hand side, making it rather like a race-course. Down the straight come these first shots.

> Swift as a thought the ball obedient flies,
> Sings high in air and seems to cleave the skies,

or else,

> Along the green the ball confounded scours,
> No lofty flight the illsped stroke impowers.

I feel that only verse should describe this initial blow, and what better than to extract from Thomas Mathison? Already the next party has moved into position:

> Ardent they grasp the ball-compelling clubs,
> And stretch their arms t'attack the little globes.

What is so marvellous about St Andrews is that anyone can play upon any of its four courses. The Old Course is *not* exclusive; indeed golf upon it is non-stop in the season, and about eighty thousand rounds are played upon it in a year!

In July, August, and September you must ballot for a place on the tee the day before you intend to play. You put your name, and the time you would like to have, in the boxes (anyone in St Andrews will explain this to you), a draw is made and times posted at various vantage-points in the town. If you had hoped for 9.50 and you get 11.45, well—you want to play? So you arrange your day accordingly and at the allotted hour, having bought your ticket—off you go, any one of you, to play what part you can on this Hamlet of all golf links.

Make no mistake: here is greatness and mystery and that strange element which is both real magic, sleight of hand, and a brilliant

[165]

just keeping on 'the windy side of the law'. In the same way as Jupiter would descend as a swan or a bull or a shower of gold to gratify desire, or a Proteus in wrestling changed shape and form most alarmingly, so St Andrews will employ every means to deceive, flatter, cajole, or dragoon you into loving it, and into admitting its mastery of *you*. Now, nobody who plays once or twice, or twenty or thirty times, will begin to *know* this links at all. That takes years or a lifetime.

But one thing is certain: it is unique, even at a first superficial glance. For it is a unique shape; rather like lesson one in learning knots, for it has a loop at the end round the 7th, 8th, 9th, 10th, and 11th and then returns the way it came: and, apart from the 1st, the 17th, and the 9th, all the greens are enormous double greens: thus 2 and 16, 3 and 15, 4 and 14, etc., are each a huge undulating putting surface; white outward flag, red homeward, waving in parallel like two terminals on a battery. This came about because in the early nineteenth century the course had only nine greens and you putted, both out and home, into the same hole; whoever arrived at the green first having precedence. But balls flying into your face became a menace even to the tough St Andrews golfer, and by 1887 the course was roughly the shape and lay-out it is now. Incidentally, the cry of 'Fore!' so feebly and apologetically crooned on English courses, has at St Andrews its true fiery leonine value; and it has modulations elsewhere unknown. There is an intonation of Fore! to cover every contingency, and like Touchstone's seven forms of deceit, it has seven shades: there is the Fore courteous, the Fore mock-modest, the Fore churlish, the Fore valiant, the Fore quarrelsome, the Fore circumstantial, and the Fore direct. I have heard them all, the Fore direct being always in unison and terrible as the roaring of a bull.

Golf never stops on the Old Course, and lost balls must be left or you are immediately bumped. Also, golf is a very slow business between shots: a wait on every tee and for most second shots. That is, of course, in the summer season when soft is the sun. You will never have the course to yourself, but if you can brave the winter's rages you will play more quickly.

St Andrews is colourful, for there is plenty of heather and whin to contend with (though not, of course, what it was). I think it is likely to surprise the newcomer how much heather there is. But you must remember that the Old Course is right up and down the middle of the long sea-heath and has both the New, the Jubilee, and the dunes between it and the North Sea, and upon its left flank it has the Eden Course, where once reposed the Elysian fields. But never was a course so like the motions of the sea: the slow, steady start of the first four holes, the sudden stride of the long 'Hole o'Cross', this is just like the movement of the ebb—running quickly now through 'Heathery' to its farthest out, the 7th, the 'High Hole'. Now comes slack water in truth, for here is the loop and 8, 9, and 10 at flat, slack-water holes, to be exploited with all possible power. It is as if here at low water the old wreck shows its ribs, and one has just time to dig for the treasure in its hold before the tide turns again. No. 8 is short—the only outward short hole and a fairly easy 3; you can get 3's at the 9th and 10th too: and now as, just before it turns, the tide seems suddenly to recede a little farther and then begin to ripple inwards, now comes the short 11th. Its green pairs with the 7th but is a little farther out on the very edge of the links, so that a shot over the green seems as if it will drop splash! into the estuary of the Eden (and it nearly will; but, instead, will lodge in a hell of sand and marram grass). To guard the green—this end of it is tiny—there are two terribly cunning bunkers, 'Strath' for a slice and the 'Hill' for a pull. Get a 3 here and it is a pure gold doubloon. But Long John Silver's parrot may easily mock you with its cry, 'Pieces of eight! Pieces of eight!' You *can* steal four 3's round the loop, you *should* take 3, 4, 4, 3, but . . . the incoming tide may float pieces of torn cards in with it.

From the twelfth tee to the eighteenth green, in floods the tide, and perhaps you are being lucky enough and playing well enough to come in with it till you reach the 'Road Hole', where the last sandcastle of your golfing pride and hope must stand against the waves until you are safely on the eighteenth tee. Many commentators have written about What Happens at the Road Hole in as erudite a fashion as Professor Dover Wilson's *What Happens in Hamlet*; yet like that great work of genius it survives all attempts to explain it. Mr T. S. Eliot has said that *Hamlet*, as a play, is an artistic failure. Perhaps the Road Hole is, too.

But confronted with the magic of Old Course, and now with those black sheds that beetle o'er their base upon the tee, we are at a climax. The courageous or foolhardy cut across over them, the staid play safely leftward. Now the Road Hole is so called because exactly along the back edge of the narrow green, down a grassy sheer drop of three feet or so, *is* a road. It is necessary to remember that this road is *below* the level of the green; because it means that the recovery shot from the road surface, which is rough, bricky and cindery, and altogether horrible, must be *pitched* up on to the sliver of green and against the slope of it. What makes the player fall into this abyss? In the face of the green just left of centre is the Road bunker, so perfectly placed that it dominates play even from the tee, and by sheer force of its spidery personality drives its victims either to avoid it too carefully and chance the road, or play too safely, and so come into its parlour, and then, of course, a bunker shot just too well out will go into the road anyway. But what is the devil in the hole is its perfect length—you *can* get on in 2—you badly want a 4—and . . . but I am presuming upon too brief an acquaintance. I have no doubt all kinds of other subtle horrors have escaped my notice. But I must record the nice, mordant humour of another of its bunkers which lies between the 'Scholar's' bunker and the 'Road'—it is called the 'Progressing' bunker.

There are many named bunkers on the Old Course whose separate histories I wish I had time to discover, and relate. You must play many times before you can relish the real nicety of being in the 'Pulpit' or the 'Shell', the 'Cat's Trap' or the 'Lion's Mouth'; but coming to the 14th, with its famous bunker, I could not help recalling Dr Faustus's question to Mephistophilis as to where Hell was, and Mephistophilis' answer:

> Why this is hell, nor am I out of it. . . .

And it is strange to think that golf was being played over this ground when Marlowe's superb play first shocked and delighted Elizabethan audiences.

The decade of the 1890's was indeed the great decade of golf, as of so much else in English life. What a ten years! Taylor, Braid, and Vardon . . . Ibsen, Shaw, and Wilde . . . in 1897 the New Course was opened to relieve pressure on the Old. The New Course, next to the Old, is over the same kind of country of heather, whins, and dune, yet it is very, very different, and I do not think it should ever be compared with the Old. It is long, difficult and, yes, dull. Good, solid, testing golf, but I'm afraid I must repeat, a bit boring. I apologise to its staunch supporters and again I know that I do not know it well enough, but I feel that there is not very much to know. Pick it up and plump it on the other side of the bay by Leuchars aerodrome, and I rather doubt if it would be more than a longish holiday course, but I certainly recommend it as such.

In that year '97, too, the Royal and Ancient finally came to dominate, in fact, a world which it had ruled by right from time immemorial, for the Rules Committee was formed to legislate for the whole game, as it has done ever since. It was incidentally William IV, the Sailor King, who graciously permitted the title of 'Royal and Ancient'; and let us not forget that our present King is a past Captain of the club.

The William IV medal, that is the autumn medal, is perhaps the club's most treasured trophy, and one story concerning it cannot be left out.

Medal day in 1860 was a day of appalling gale and just as a certain Captain Maitland-Dougall, R. N., was about to start there was a cry of 'A wreck! A wreck! Man the Life-boat!' The life-boat was launched, but there were not enough volunteers. The gallant Captain downed his clubs, jumped aboard, and stroked the boat to a successful rescue. At the end of which, hours later, wet, sopping, and one would have thought exhausted, he returned to the tee—and won the Medal!

But stories of golfing feats at St Andrews are legion, and that one must stand for all of them; legion also are the names of the heroes: Old Tom and Young Tom, Allan Robertson, Andrew Kirkaldy—it needs a Homeric touch to list them, and their deeds

There is now a 'limited Sunday service' of golf at St Andrews, play being permitted upon the Eden Course, which, in character, is as far a cry from its neighbour as a play by J. B. Priestley is from Shakespeare. But, in general, Sunday is a day for meditation—perhaps upon Mr Eliot's summing up of the century we live in:

> Their only monument the asphalt road
> And a thousand lost golf balls.

Perhaps it is a day for a little quiet putting on the drawing-room carpet, or for finding some of those errant thousand, or even for reading books about golf . . . certainly it is a day of expectancy; a day on which—since we may not play on the Old Course—we are all scratch players on it:

> Thou giv'st me to the world's last hour
> A golfer's fame divine;

I boast—thy gift—a Driver's power,
If I can Putt—'tis thine.

PATRIC DICKINSON, *A Round of Golf Courses*, 1951

PINE VALLEY

There is a sense of privilege as well as rare experience in visiting Pine Valley, for it has no parallel anywhere. No course presents more vividly and more severely the basic challenge of golf—the balance between fear and courage. Nowhere is the brave and beautiful shot rewarded so splendidly in comparison to the weak and faltering; nowhere is there such a terrible contrast between reward and punishment, and yet, withal, the examination is just. The fairways are not narrow and the greens, which demand long approach shots—are large and welcoming. It is not necessary to be exceptionally long or accurate, but it is essential to be able to hit the ball reasonably well. If the fairways and greens are missed then punishment can be terrible indeed, because the ground about them has been left in its natural state. There is no rough to speak of, for the woods border every hole; but the enormous bunkers are never raked and water comes terrifyingly into play at four holes. Here, surely, is tradition rooted in the first principles of the game, for golfers of olden times were not afraid to play the ball as it lay.

Successful scrambling is quite out of the question. It is impossible to hit the ball badly and score reasonably, and precious few courses, if any, can justly claim likewise. Therefore Pine Valley is not for the meek and frail though thousands come to try their hands, some to curse and some to glory, almost masochistically, in their high scores. Here no one is ashamed of his total. Instead, finishing the course is a matter of pride, for the members well know how easy it is to take an awful number of shots awfully quickly.

When you first play Pine Valley, imagination is tuned to its most destructive pitch. You have heard all the gruesome tales of men taking 27s and the like, of ghastly failures on the brink of success, of the wagers that are laid that the ordinary player will not break 100 for his first round and that receiving five strokes a hole you will not finish eighteen up on bogey. And so the holes look narrower than they are, the carries longer, the trees more menacing, and the water, which must be carried, desperately wide. And yet no hole is beyond the scope of the average golfer, provided he uses and keeps his head. But once let him stray, then sevens and eights and far worse will crowd thick and fast. The great hazard of Pine Valley is psychological. If one can forget its reputation and overcome the fear of its appearance, then half the battle is won. But therein lies a greater part of the test of golf.

The course is a masterpiece of architecture, imagination, and beauty and the temptation to describe every hole is great. Of all the courses I have seen none remains so vividly in the mind. Each hole is different, setting new problems of judgement, control, and planning. Many are constructed on the island principle with fairways and greens as oases in a desert of scrub, woods, and sand. Such is the second with its sloping green a pyramid amidst desolation and the seventh, one of the classic long holes in the world. If the drive is good, then an attempt may be made to carry a hundred yards of wilderness; if it is not, then it is advisable to play short. The green, too, is surrounded with sand and no one has or ever will reach it in two.

The short holes are magnificent, even if at first they strike terror to the heart of the bravest. The third, a long iron downhill to a lonely, beautiful island green, and the fifth offer no compromise whatsoever. Thinking of the fifth makes me shudder even now. A fair stretch of water lies below the tee and a carry of almost 200 yards to a green tucked high in the woods must be made. All one had to do was to hit a driver shot pretty straight into the wind but alas . . . The tenth, a short pitch, is easy if the shot is right; if not, the ball may finish in an unprintable part of the devil's anatomy as the bunker beneath the green is locally described. The last short hole is the most famous. Its green lies on a neck of land far below the tee with water on three sides where turtles are wont to play and which once yielded 15,000 balls. Perfectly simple if a medium iron is hit properly. This same water stretches for 150 yards in front of the fifteenth tee. You stand there until you make it and then walk through the woods to one of the longest fairways in existence which grows narrower after a quarter of a mile.

All, however, is not terrifying. Far from it. Given a straight drive, the huge fourth green positively begs hitting; the eighth may bring a birdie if one does not socket a tiny pitch to a tiny green, as may the twelfth where the fairway should be large enough for anyone. The thirteenth is a classic par four. After a good drive to a plateau the hole swings downhill with all kinds of perdition on the left which cut into the line of the second shot. But like all great holes there is an alternative. Safety and a probable five lie to the right. The sixteenth is by no means alarming, once the tee shot is away; neither is the seventeenth, if one ignores what has to be carried with the pitch to the green. It is said that one golfer once took 67 there but he probably lost his temper. I recall blasting from a cactus plant after just missing this fairway.

From the eighteenth tee the towers of Philadelphia can be seen in the shining distance. The fairway tilts towards the woods far below and sanctuary is near, but first a brave shot must be hit from a sloping lie across a stream. And then Pine Valley has been played and a lifetime's memory made. There remain only post-mortems, the like of which are heard nowhere else, in a clubhouse as cool, shadowed, masculine, and peaceful as any Englishman could desire, and infinitely welcoming. A portrait of Crump glows in the darkness and his spirit lives on in men like John Arthur Brown, president these thirty years: men who abhor the modern urge to simplify, who love golf because its challenge is undying, and who are proud of their privilege to preserve it at Pine Valley.

PAT WARD-THOMAS, *Masters of Golf*, 1961

THE MAKING OF AUGUSTA NATIONAL
I shall never forget my first visit to the property which is now the Augusta National. The long lane of magnolias through which we approached was beautiful. The old manor house with its cupola and walls of masonry two feet thick was charming. The rare trees and shrubs of the old nursery were enchanting. But when I walked out on the grass terrace under the big trees behind the house and looked down over the property, the experience was unforgettable. It seemed that this land had been lying here for years just waiting for someone to lay a golf course upon it. Indeed, it even looked as though it were already a golf course, and I am sure that one standing today where I stood on this first visit, on the terrace overlooking the practice putting green, sees the property almost exactly as I saw it then. The grass of the fairways and greens is greener, of course, and some of the pines are a bit larger, but the broad expanse of the main body of the property lay at my feet then just as it does now.

I still like to sit on this terrace, and can do so for hours at a time, enjoying the beauty of this panorama.

With this sort of land, of a soft, gentle, rather than spectacular beauty, it was especially appropriate that we chose Dr Alister MacKenzie to design our course. For it was essential to our requirements that we build a course within the capacity of the average golfer to enjoy. This did not mean that the design should be insipid, for our players were expected to be sophisticated. They would demand interesting, lively golf, but would not long endure a course which kept them constantly straining for distance and playing out of sand.

There was much conversation at the time to the effect that MacKenzie and I expected to reproduce in their entirety holes of famous courses around the world where I had played in competitions. This was, at best, a bit naïve, because to do such a thing, we would have had literally to alter the face of the earth. It was to be expected, of course, that the new lay-out would be strongly influenced by holes which either MacKenzie or I had admired, but it was only possible that we should have certain features of these holes in mind and attempt to adapt them to the terrain with which we were working.

I think MacKenzie and I managed to work as a completely sympathetic team. Of course, there was never any question that he was the architect and I his advisor and consultant. No man learns to design a golf course simply by playing golf, no matter how well. But it happened that both of us were extravagant admirers of the Old Course at St Andrews and we both desired as much as possible to simulate seaside conditions insofar as the differences in turf and terrain would allow.

MacKenzie was very fond of expressing his creed as a golf-course architect by saying that he tried to build courses for the 'most enjoyment for the greatest number'. This happened to coincide completely with my own view. It seemed to me that too many courses I had seen had been constructed with an eye to difficulty alone, and that in the effort to construct an exacting course which would thwart the expert, the average golfer who paid the bills was entirely overlooked. Too often, the worth of a layout seemed to be measured by how successfully it had withstood the efforts of professionals to better its par or to lower its record.

The first purpose of any golf course should be to give pleasure, and that to the greatest possible number of players, without respect to their capabilities. As far as possible, there should be presented to each golfer an interesting problem which will test him without being so impossibly difficult that he will have little chance of success. There must be something to do, but that something must always be within the realm of reasonable accomplishment.

From the standpoint of the inexpert player, there is nothing so disheartening as the appearance of a carry which is beyond his best effort and which offers no alternative route. In such a situation there is nothing for the golfer to do, for he is given no opportunity to overcome his deficiency in length by either accuracy or judgment.

With respect to the employment of hazards off the tee and through the green, the doctor and I agreed that two things were essential. First, there must be a way around for those unwilling to attempt the carry; and second, there must be a definite reward awaiting the man who makes it. Without the alternative route the situation is unfair. Without the reward it is meaningless.

There are two ways of widening the gap between a good tee shot and a bad one. One is to inflict a severe and immediate punishment on a bad shot, to place its perpetrator in a bunker or in some other trouble which will demand the sacrifice of a stroke in recovering. The other is to reward the good shot by making the second shot simpler in proportion to the excellence of the first. The reward may be of any nature, but it is more commonly one of four—a better view of the green, an easier angle from which to attack a slope, an open approach past guarding hazards, or even a better run to the tee shot itself. But the elimination of purely punitive hazards provides an opportunity for the player to retrieve his situation by an exceptional second shot.

A course which is constructed with these principles in view must be interesting, because it will offer problems which a man may tackle, according to his ability. It will never become hopeless for the duffer, nor fail to concern and interest the expert. And it will be found, like the Old Course at St Andrews, to become more delightful the more it is studied and played.

We try very hard in Augusta to avoid placing meaningless bunkers on the course. Some of the natural hazards are severe, but usually so for the ambitious player. Possibly the dearth of bunkers on the course is a feature most commented upon by visitors. Yet there are some which perhaps could be dispensed with, except that they are in use to protect players from more dire consequences. Occasionally a bunker may be used to stop a ball from running into a hazard of a more serious nature.

THE COURSE

Our over-all aim at the Augusta National has been to provide a golf course of considerable natural beauty, relatively easy for the average golfer to play, and at the same time testing for the expert player striving to better par figures. We hope to make bogies easy if frankly sought, pars readily obtainable by standard good play, and birdies, except on the par fives, dearly bought. Obviously, with a course as wide-open as needed to accommodate the average golfer, we can only tighten it up by increasing the difficulty of play around the hole. This we attempt to do during the tournament by placing the flags in more difficult and exacting positions and by increasing the speed of the greens. Additionally, we try to maintain our greens of such a firmness that they will only hold a well-played shot, and not a ball that has been hit without the backspin reasonably to be expected, considering the length of the shot.

Generally speaking, the greens at Augusta are quite large, rolling, and with carefully contrived undulations, the effect of which is magnified as the speed is increased. We are quite willing to have low scores made during the tournament. It is not our intention to rig the golf course so as to make it tricky. It is our feeling that there is something wrong with a golf course which will not yield a score in the sixties to a player who has played well enough to deserve it.

On the other hand, we do not believe that birdies should be made too easily. We think that to play two good shots to a par four hole and then to hole a ten-foot putt on a dead-level green is not enough. If the player is to beat par, we should like to ask him to hit a truly fine second shot right up against the flag or to hole a putt of more than a little difficulty. We therefore place the holes on tournament days in such locations on the greens as to require a really fine shot in order to get close. With the greens fast and undulating, the putts from medium distances are difficult and the player who leaves his ball on the outer reaches has a real problem to get down in par figures.

The contours of the greens at Augusta have been very carefully designed. We try to provide on each green at least four areas which we describe as 'pin locations'. This does not mean that the pin is always placed in one very definite spot within these areas, but each area provides an opportunity for cutting the hole in a spot where the contours are very gentle for a radius of four or five feet all around.

The selection of the pin area and the exact location of the hole are decided on the morning of play by a committee appointed for the purpose. The decision will be affected by the condition of the putting surface itself, the state of the weather to be expected, and the holding qualities of the ground. Naturally, the job of placing the holes on tournament days is one which calls for a considerable knowledge of the game and good judgment. I am sure that the players involved would be interested, too, that the committee be composed of individuals of benign and charitable natures.

BOBBY JONES, *Golf is my Game*, 1959

BALLYBUNION

After this piece was written in 1971, coastal erosion claimed the second green and forced a major replanning of the course. This description is therefore no guide to it in its present state. It is, however, a marvellously warm and insightful appreciation of the layout and its very special character.

A tourist driving through Switzerland is staggered by its prodigal beauty; around the bend from the most wondrous view he has ever beheld he comes upon a view that surpasses it—and so on and on, endlessly. Ballybunion is something like that. I do not mean to suggest that there are vistas that put the one from the second tee to shame—there aren't—but there is a correspondence in the way one stirring hole is followed by another and another. The third, for example, is a 145-yarder that moves through the sand hills to a devilish little green that tips into an abrupt downslope on the right and is bunkered on the left and in front. On the fourth, a 451-yard par 4 on the inland side of the sand-hill belt, the key hazard is a deep bunker carved in the face of a rise in the middle of the fairway about twenty yards in front of the plateau green; anything less than a perfectly struck second shot will end up in it. (A sign at the edge of the bunker informs the golfer that it is called the Crow's Nest. There are similar signs at the other major hazards on the

[176]

course—an original touch and a flavorsome one.) The fifth hole, which curves back toward the ocean through another twisting valley, is a very attractive drive-and-pitch par 4,343 yards long, a dogleg to the left this time. The sixth, which I've already mentioned, is a perfect beauty—a 450-yard par 4 that tumbles downhill along the cliffs to an inviting green. And that is the way it keeps going at Ballybunion. There is not one prosaic hole—not one single 'breather'—in the whole eighteen. If the course has a weakness, I suppose it is the comparative plainness of the last two holes—a pair of par 5s stretching over the somewhat featureless interior ground. They are not trite holes, for they are imaginatively bunkered, but they do lack the beauty of the rest of the course, and as a result they are somewhat anticlimactic.

At the conclusion of our walk, I found I could remember each of the eighteen holes without much trouble. This is probably the oldest and soundest rule of thumb for judging the merits of a course that a golfer has just seen for the first time. Apart from the first six holes, three others remained especially distinct: the tenth, a 210-yarder over difficult duneland, where the prevailing wind, from the west, sweeps across from left to right; the 368-yard thirteenth, where the sand in a fifty-yard-wide bunker called the Sahara is strewn with deer bones, shells, stones, and ashes deposited in the fifteenth century by a tribe that used the cavity as a midden, or dump; and the fourteenth, a 376-yard par 4 that Simpson enlivened by placing in the drive zone a mounded double bunker, which the members immediately dubbed Mrs Simpson. Ballybunion is a moderate 6,317 yards in length, par is 71 (34–37), the course record is 65, and, as I say, no other links, in my opinion, presents such a satisfying adventure in golf. The course has two endowments that I believe to be unique. First, it is the only links I know of where most of the oceanside holes are perched atop spectacular cliffs, in the manner of Pebble Beach (which is not a true links). Second—and this is undoubtedly the secret of its character and charm—it is the only links I know of where the sand-hill ridges do not run parallel to the shore but at a decided traverse. This opens all sorts of possibilities—dogleg holes of every description sculptured through the choppy land, and straightaway holes where the sand hills patrol the entrance to the green like the Pillars of Hercules. One more point: Unlike most links, Ballybunion challenges you with target golf. There is none of that bouncing your iron approach short of the green and letting it bobble toward the flag, as you do on most courses in the British Isles. No, you aim for the flag, and if your shot lands on the green the green is sufficiently receptive to hold it.

HERBERT WARREN WIND, *Following Through*, 1985

PEBBLE BEACH

Next morning . . . we had a tilt at Pebble Beach. With one or two unforgettable exceptions the general scene is not quite so spectacular as Cypress Point but from the back tees it would, I think, with even greater certainty crush an indifferent golfer into total oblivion. As a generality I should say that American courses are less difficult than British—partly because the greens are softer and you can therefore play 'target golf' most of the time but mainly because in the golfing season (and courses in the north are closed in the winter) there is little or no wind. Pebble Beach, hard on the shore of the ocean, is among the exceptions and the wind can blow in with immense ferocity. I am aware that the wind also blows in Texas and this, according to Hogan, explains why the best golfers like Demaret, Mangrum, Burke and, though he does not say so, himself, as well as the best and biggest of everything else in the United States, come from Texas. For most American players, however, golf is a still-air game.

Pebble Beach reminded me in many ways of a longer, sunnier, and more sophisticated section of that lovely, all too little known course of the west coast of Ireland—Ballybunion. In the technical sense I suppose there is no difference between driving out of bounds into the Field at Hoylake or over the little stone wall at St Andrews or into the Bristol Channel at Porthcawl or, as I habitually did for the first dozen years of my golfing life, the cornfield, the allotments, the Great Ouse, or British Railways (Midland Region), which bound the course at Bedford. Yet, psychologically, there is all the difference between driving into these and driving into the Ocean. Oceans are hazards on the grand scale. They make one's efforts so much more puny and insignificant than mere seas or rivers. At Ballybunion, in winter one of the wildest places in the British Isles, with the original Cruel Sea pounding away on the rocks, a high slice is swallowed by the Atlantic like a grain of sand. A big enough slice theoretically would see no land till it touched Long Island. At Pebble Beach a good hook from the 18th tee, missing the tiny atolls that dot the surface of the Pacific, would presumably fetch up in Japan.

For five holes, as a matter of fact, Pebble Beach reveals little of what is in store. Wandering soberly among the fir trees it leaves one wondering politely what all the fuss was about. One's score at this moment may well, though mine wasn't, be one under fours. At the sixth it comes out of its corner, as it were, and delivers four tremendous body blows in succession. The sixth, after a comparatively innocuous drive, leaves you playing a brassie to a green set high up on a headland, over which the wind may be tearing with frenzy. Eating into the fairway on the right is what in Devonshire would be known as a 'combe', thick with undergrowth and leading to the rocks. A ball slightly 'skied' is caught by the wind and whirled away like a feather.

Across the lower slope of this headland on the far, or ocean, side is a fascinating little hole of only 110 yards, the shortest perhaps on any course of this calibre, but, whereas on most country club courses this would be the merest 'pushover', here on the edge of the Pacific, with the spray forming a dramatic background to a tiny green surrounded almost entirely by sand, it can be anything up to a No. 3 iron.

Having got past this, you then receive the full impact of this essential quality of Pebble Beach, its power to surprise. You drive uphill at the eighth, along the side of the promontory opposite the sixth—a most commonplace stroke with nothing apparently to play for except to move the ball along and keep it in play. Having done this, you trudge up the slope and reach the crest, where with any luck your ball is lying.

'Good God!' I remember saying. 'Look at that!'

A few yards ahead the fairway ceased abruptly at the edge of a cliff perhaps 150 feet high. The cliff worked its way inland to the left, forming a bay perhaps 180 yards across as it curved away back again across our front. Directly ahead on the far side of this bay was the green—tightly trapped, for good measure, with sand bunkers. If ever there was a death-or-glory shot, this was it—and a full shot with a brassie at that. I contributed a couple of balls to the ocean— it is almost impossible not to look up too soon, in order not to miss a moment of the ball's spectacular flight across the bay, the result of which is of course, that, half topped, it disappears instantly from sight—but my third was a beauty and I can see it now, soaring over the chasm, white against the blue Californian sky, and pitching with a thump on the green.

Innumerable people must have torn up their card at the ninth. Many, on looking back at it, must have recollected the same sequence—lulled into false security by the first five, softened up by the sixth and seventh, sent tottering by the eighth, and knocked out by the ninth. Even the great Byron Nelson was once battered into submission by the ninth and gave in after about 11 shots. It is a long hole, parallel with the beach, with the tiniest opening to the green for those who go for it with their second—on the left a sort of chasm, on the right our old friend and enemy the Pacific.

After this comes a mild respite, enlivened, as I have mentioned, by the prospect of being able to go home and say you have sliced into Bing Crosby's garden—I imagine that he never has to buy a ball, being supported wholly by involuntary contributions—but you emerge from the trees again with a fine one-shotter into the teeth of the wind at the 17th, with the green on the edge of the beach, and then you work your way home, crescent-wise, along the 18th, which I believe I am right in saying no one has yet reached in two. After which you ascend, bloody but unbowed, to further gossip with Peter Hay in a 19th hole which is adorned once again with pictures of the Masters old and new—Tom Morris, Harry Vardon, Francis Ouimet, Hagen, Jones, and the rest who have helped to turn golf into the world-wide language that it is.

HENRY LONGHURST, *Round in Sixty-Eight*, 1953

THE ROUGH AT PEBBLE BEACH

Pebble Beach is one of the most beautiful places in the world to play golf, albeit attended by one of the world's worst golfing hazards, the *mesembryanthemum nodiflorum*, otherwise known as ice plant. There are times when ice plant can look like a nice plant—it blooms prettily in purple and pink—but those times are not when one's ball is in it.

Because the plant helps to prevent erosion it is valued on coastal properties. It is abundant on Pebble and Spyglass Hill and is the cause of more high scores and lost tempers than all the other hazards put together. It grows in thick green fingers, irregularly spaced, and its particular frustration is that a ball on it is not only visible but always looks playable.

On swinging a club, though, one finds it is like swinging at a rubber tyre. The plant has a resistant resilience that defies passage to the ball to all but the best, or most vicious, of swings.

J. C. Snead, nephew of Sam, tells of how they were playing Spyglass one day and Sam's ball finished in the ice plant at the back of the short 5th, sitting up. 'I said to him, "Unkie, ever play out of that stuff?" He said "No" and I said, "Better just take a drop out of there." He gives me a double take like "Boy, do you know who you're talking to?" and takes his eight-iron. He does his little waggle, gives it a flick and the ball hasn't moved.

'He ends up with a six. Then he birdies three of the last four holes and I say to him: "Unkie, a bogey on that par-three would look pretty good right now, wouldn't it?" and he looks at me and says "Damn if it wouldn't".'   DAVID DAVIES, *Guardian*, 1992

❖

OAKMONT

. . . Oakmont stands as the avatar of the ultra-stringent qualities that the United States Golf Association has always set such store by in choosing and preparing the venue of the national championship. To begin with, it is a long course. (As far back as the 1927 Open, it measured over 6,900 yards—an almost unreasonable length in those days.) Its fairways are kept so narrow for everyday play that this year the U.S.G.A., which always makes it a point to bring the rough in drastically on its Open courses, so that the fairways seldom exceed thirty-five yards in width, had to ask Oakmont to widen two of its fairways. On top of this, Oakmont boasts the fastest greens in America, which also happen to be among the truest. For several Opens, they have been cut to two-thirty-seconds of an inch—three times as short as is standard for the Open. Put it all together and you can understand why Walter Hagen spoke of Oakmont as 'the ideal championship spot,' and why Tommy Armour called it 'the final degree in the college of golf.'

As the old saying has it, behind every great golf course is a great man. The man behind Oakmont was Henry C. Fownes—pronounced 'phones'—a pioneer Pittsburgh steel magnate, who in 1899, when he was already in his forties, decided to take up golf, a game that his fellow steel-maker Andrew Carnegie once described as 'an indispensable adjunct of high civilization.' Fownes, a short, compact man, first played his golf on the six holes of the Pittsburgh

Field Club, but he switched the next year to the nine-hole Highland Country Club. In a short time, he became a surprisingly able player, particularly skillful around and on the greens. In 1901, he qualified for the United States Amateur and won his first match. He qualified for the Amateur in 1902 and again in 1903, when he won his first three matches. In 1907, when he was in his fifties, he qualified for the Amateur for the last time. There is no record that Fownes travelled to Britain around the turn of the century to play and study the classic courses of Scotland and England, but he obviously read a good deal about them in books and magazines. In 1903, when he made up his mind that the time had come for him to organize a syndicate and build a modern eighteen-hole course, there was no question whatever as to what his goal was: if Pittsburgh's distance from the ocean denied him the opportunity to build an honest-to-goodness British linksland course, he would jolly well build the next best thing possible—a real British moorland course. Assisted by his son, William C., Jr., who had been named for his uncle but was oddly given the 'Junior' anyway, Fownes reversed the usual process. He first laid out on paper eighteen sturdy holes and then set out to find a suitable tract of land to build them on. He had been looking for a while when a friend of his, George S. Macrum, who lived in the village of Oakmont, some twelve miles northwest of downtown Pittsburgh, called his attention to a two-hundred-and-twenty-one-acre sweep of gently rolling farmland in that area. After inspecting it and consulting with the members of his syndicate, Fownes purchased the property. At seven o'clock on the morning of September 15, 1903, under the personal direction of the Fowneses, *père et fils*, a work force of a hundred and fifty men, along with twenty-five teams of mules, began construction of the course. Six weeks later, before wintry weather set in, the first twelve holes were completed—the tees and greens built, the drainage ditches dug, and the fairways and greens seeded, the latter with a mixture of South German bent. The next spring, the remaining six holes were finished, and that autumn Oakmont was opened for play. It was not a course to regale the eye. Fownes had had his crew chop down just about every tree in sight except for a few around the gabled clubhouse, and this, coupled with the flattish terrain over which the holes were routed, did indeed give Oakmont the not easily achieved British bleakness that Fownes was after. It was a terribly long course from the start—6,600 yards from the back of the tees—and saturated with bunkers. Fownes left no doubt about his philosophy of golf. When a man did not hit an almost perfect shot, he was supposed to pay a stiff penalty for his error. From the outset, American golfers agreed on two things about Oakmont: if it was the most punishing course in the country, it was also the best conditioned. Under the Fownes family, it was treated like a living thing. As early as 1906, the club sank ten thousand dollars a year into it. As maintenance costs rose, so did the money set aside for its maintenance. During the Depression years, the Fowneses personally picked up all the bills.

The first national championship held at Oakmont was the 1919 Amateur—Davy Herron, an Oakmont member, defeated seventeen-year-old Bobby Jones in the final—but it was the 1927 U.S. Open that made the golf world really conscious of the course. H. C. Fownes, the club's first president, still held that office, as he would until his death, late in 1935, but he was getting along in years, and W. C. was now running things. Although he was neither as imperious nor as stubborn as his father, W. C. was far from reticent when it came to expressing himself on golfing matters, and with some reason. He was a talented enough player to win the Amateur in 1910, and he qualified for that championship twenty-five times in twenty-seven attempts—a terrific record. He made it a point to know the game from all sides, and after a long administrative career with the U.S.G.A. he served as its president in 1926 and 1927. Where Oakmont was concerned, as the perennial chairman of the green committee he meant to keep it a sanctuary of par in a world gone mad with birdies and eagles, 67s and 65s. Toward this end, beginning in 1920, the bunkers at Oakmont were furrowed—that is, Emil Loeffler, the greenkeeper, combed them with a heavily weighted metal rake whose triangular teeth were two inches long and set two inches apart. Oakmont's bunkers in those days were filled with a coarse brownish sand that came from the Allegheny River, and it had enough body to hold the creases. Exploding a ball from a furrowed greenside bunker wasn't too much of a problem—the golfer simply blasted out the ridge of sand behind the ball along with the ball—but if he caught a furrowed fairway bunker off the tee he could not hope to advance the ball farther than forty yards. A good many golfers felt that this was plain unfair, but the Fowneses had a ready rationalization. The fairways on many of the best British courses, they pointed out, were punctuated by deep, sharp-walled bunkers that restricted the length of the recovery shot. Since bunkers of comparable depth could not be installed at Oakmont—the soil was extremely clayey, and this would have presented a serious drainage problem—the only way to make the shallow bunkers a menace was to furrow them. What with its fearsome bunkers and its glassy greens (which were massaged by a fifteen-hundred-pound roller that required eight men to handle it), not to mention its sheer length (6,965 yards), it was not at all surprising that nobody in the field broke 300 at Oakmont—par 72—in the 1927 Open. Armour tied for first with Harry Cooper at 301, thirteen over par, and then beat him in the playoff with a 76. Jones, incidentally, never broke 76 on his four rounds.

For the 1935 Open, the U.S.G.A. insisted that the furrowing at the bunkers be considerably modified, but, apart from that concession, Oakmont played at least as hard as it had in 1927. In the interim between the two championships, a good many of the holes had been remodelled, and in the process the course had been extended to 6,981 yards. There were thirty more bunkers than there had been for the 1927 Open, bringing the total to well over two hundred. (The largest was the Sahara, on the short eighth—a monster seventy-five yards long, into which eleven railroad carloads of sand had been dumped.) There were also twenty-one traversing ditches to worry about. Nevertheless, the Fowneses remained eternally fretful that some golfer would come along one day and reduce to a shambles the course they wanted to be accepted as the best test in the world. Gene Sarazen loves to recall a fascinating incident that took place about a month before the 1935 Open. One weekend when W. C. was out of town, a visiting power hitter carried a bunker on the seventh fairway—about two hundred and forty yards from the tee—that no one was supposed to carry. Loeffler, following instructions for such emergencies, immediately telephoned W. C. and reported what had taken place. W. C. did not terminate his weekend then and there and hop the next plane back, but the first thing he did when he returned on Monday was to go out to the seventh hole and, after appropriate study, order Loeffler to put in a small, flat bunker just beyond the one that had been carried. Measure for measure. In the 1935 Open, only one man broke 300. This was Sam Parks, Jr., a young pro from the Pittsburgh area, who drove well, carefully thought out his approach shots to the extremely firm greens, putted steadily, and brought in a total of

299. No one in the entire field broke 70. On the last round, none of the twenty leaders broke 75. Really!

W. C. Fownes succeeded his father as president of Oakmont in 1935, and continued to head up the club's green committee— W. C.'s first love. In 1949, the year before his death, his duties were assumed by an eighteen-man board. In November of that year, this board declared that it was well aware it had 'inherited what had become known as the finest golf course in America' and dedicated itself to perpetuating Oakmont's lofty traditions. Under the new leadership, a few bunkers were filled in and the speed of the greens was reduced a shade. As a result, the course that the field confronted in the 1953 Open was a somewhat more humane proposition. That was one of the two reasons Oakmont was at length tamed by Ben Hogan, whose winning total was 283, five strokes under par. The second reason was that Hogan seldom in his long career hit the ball with such supernal control throughout four rounds. In this general connection, a few weeks ago, before the recent Open, a visitor to Oakmont happened to ask Frank Ingersoll, a contemporary and close friend of W. C.'s, what would have been Fownes' reaction had he been around to watch Hogan dominate the course so thoroughly. Would he have saluted Hogan for the brilliance of his shotmaking or would he have been chagrined that par had finally been flaunted? 'Bill Fownes was in the habit, as you may know, of going out onto the course at regular intervals and checking on how the different holes were playing,' Ingersoll said, in a slow, thin voice. 'He'd go out to some hole and sit there on his shooting stick for three or four days. If he saw that a lot of drives were ending up safely in a certain spot, he'd do something to tighten up the shot—put in a new trap or extend the rough or enlarge an old trap. No, I don't think he would have been delighted by Hogan's performance. No, I think he would have probably said to himself, "It's your own fault, Fownes. If you had spent more time out on the course studying how the holes were playing, this never would have happened."'

HERBERT WARREN WIND, *Following Through*, 1985

❖

ROYAL PORTRUSH

Portrush is H. S. Colt's masterpiece (though he may not agree with me). Certainly the ground is worthy of a great links; there is spaciousness and grandeur; these ranges of sandhills are, to ordinary sandhills, what the Alps are to the Grampians. Out to sea lie the rocky ridges of the Skerries—originally, I imagine, the coastline, for either side of the links the cliffs soon rise again dour and grey: but within the Skerries the Atlantic has swelled and broken, and piled in rollers of sand, creating a stretch of country on the grand scale. The sandhills rise shelf upon shelf, and it is upon the most landward shelf that the links is set. One has the curious feeling of playing along a cliff top—high above the sea, but the whole *cliff* is made of sandhills. I have never seen a links which so invites adjectives of nobility and size; of space and height. Sometimes in dreams one has the sensation of flying or gliding gently and easily as a gull from the cliff-side. Portrush is flying golf—one longs to take off after the ball, and indeed the air from the Atlantic is so fresh that such levitation does not seem impossible.

The rough is composed of wild briar rose, thick, long claggy grasses and (if there is a wet season) moss; the whole compound has the consistency of porridge and is entirely hellish; a kind of Sargasso. I speak first of the rough because Portrush is proud—and rightly proud—of the fact that it has very few bunkers. (You may remember that another giant, Carnoustie, is busy removing its surplus.)

You will not be impressed by the first hole, though you may be irritated. How wide the fairway, which slopes gently down and away with a rather meadowy inland look; how well to the left that boundary fence; how far to the right that other fence; yet in the days of stroke and distance a scratch player has been known to play 13 from the tee—that means six consecutive shots out of bounds. It is so *unlike* the first drive at Hoylake, so charmingly obvious and avoidable, and yet . . . A straight long drive and all is well. I think Portrush is particularly a driver's links. There is trouble in front of most tees, real ghastly give-it-up-at-once kind of trouble; but the carries—even from the back tees—are not fierce. Anything crooked goes into the porridge—there's no chance of landing on another fairway and getting away with a long, wild wicked one. No, the ball from the tee must be hit straight and true. That, I know, is one of the oldest clichés of the game; at Portrush you'd better forget it's a cliché and come freshly upon it as a great Discovery: the one magical secret of golf! Note that I have just put the adjectives 'straight' and 'long' in that order. Second shots are a joy, there are plenty of opportunities to use wood; and where that is so, you are given that special kind of do-or-dare shot which *may* reach the green—yet, if hit correctly, gives you entire pleasure even if you do not. Such a heavenly swoop is the second shot at No. 2. Away goes the ball downhill (so to speak) through a curving pass, or defile, just out of sight, to reappear at rest across a further gully and on the edge of the green. There are lots of chances for the pitch-and-run from twenty yards—owing to this lack of bunkers—for Colt has placed his greens brilliantly, achieving here, I think, the perfect mating of the old-fashioned cup-in-a-hollow and the new-fashioned mushroom-on-a-hill style of green. Owing to the largesse of Nature he has been able to place his greens *up* and still to have them among hills whose shoulders and slopes diminish their size to pinheads. This makes the long second, or the short third, always an exciting and exhilarating shot.

From the second green you climb steeply up to the third tee and from it you have one of the most marvellous views: landscape and seascape and golfscape. There is something quite ravishing to the golfer in these winding viridian ways towards the emerald with the bright flags stiff in the wind; the 'silence and slow time of it' and the pacing, intent figures. And on the third tee, here, the tract of sandhills appears so vast, the human figures so small that one looks with a hawk's or an airman's eye, and having spent the lyrical moment, like stout Cortez and all his men, the wonder gives place to the wild surmise: how long this shaft is! how small this clubhead, how minute this ball! How far the green! *How can I do this at all?* Luckily for you the 3rd is a short hole, and not a very difficult one. It is called 'Islay' and there *is* Islay out on the horizon. But on, to the fourth hole—a beautiful two-and-a-half shotter like the word 'every', which you can give three or two syllables to, as you choose. There is an out-of-bounds fence along the right and a well-placed fairway bunker nudging you over towards it, and another of those long hopeful seconds taking you to the mouth of a green beautifully placed in among dunes.

The greens are, on the whole, easy to putt on. They are not over-large and not too full of borrows. One should never get the horrors on them. They are greens for the attacking putter.

Now for the 5th: one of the most romantic holes in the world. (And this is a great romantic links.) Standing on the highest point of the dunes inland (and we have come east all along the inside

perimeter between dunes and wild hill-side), you turn at right angles
and face directly seaward. The ground collapses down towards the
sea in a huge series of lumpy mounds, a steeply sloping field of
giant molehills; the gradient is violent and away below and to the
right, on the edge above the sea, is the green. It looks miles away.
But as the ball flies it is only a drive and a No. 3 or 4. This drive
is really enchanting. You must hit well to the left of where you'd
like to—for over the hills and less far away, there is a fairway which
you can't quite see from the tee. The more you bite off on the
direct line the more likely you are to land among the briary hills
and to take 3 more (the 1st backwards and the 3rd sideways) to
reach the fairway. I cannot begin to describe the fantasy of this
awesome and beautiful golf hole. It is like playing down a long
doglegged waterfall with the green a still, deep, green pool below
the rapids. And if you go over *that*, you go down, down, into the
sea. It is well that one turns inland to the next hole and that it is
a long one-shot hole of a reasonably tame nature: but only one
generation from wildness, like a tiger-cub born in a zoo. The talons
on its green paws know what to do.

Since 1945 two new holes have been made in these first nine—
and very good ones they are—in particular the two-shot 8th, a
sharp dogleg to the right whose angle is perfect, as the straight,
long, driver will find. The green is a not very long but very narrow strip between sandhills and if you are not opposite the entrance
your second will sink into the sticky and prickly obscurity which
surrounds the green.

Like the old song 'Yes, we have no bananas' it is 'Yes, we have
no sand-bunkers' . . . you may remember how the song went on to
catalogue what other fruit and vegetables were in stock. The 8th
will tell you what *it* has in stock, if you go crooked. There is a
refreshment hut at the turn; and the 10th is not too cruel, after it,
being a nice steady-going chap of a bogey-five character, and the
11th is a charmingly precipitous short hole called 'The Feather Bed'.
You just flick the ball away straight and down it drops plumb—
into one of the bunkers which draw a cordon round the green, as
if it were wanted for some crime and was about to make a bolt
for it. The 12th has a very small seductive tortoiseshell of a green
from which second shots retreat rejected in a thwarted way. I must
admit that I do not admire this hole as much as I am told I should.
I feel it has somehow had its sting removed, it feels bowdlerised.

But for the next two holes my admiration is unbounded. The 13th, like the 5th, points seaward, but this time you drive up a steady slope to the crest of a ridge. Across a small valley lies the green, again on the sea-edge with a deep rushy chasm to its left and a steep convex slope to its right—in fact it is placed exactly on the top of a large dune, and it needs a nice straight, high No. 5 into the wind or through it. This is a beautifully proportioned hole and gives no hint of what is to come.

Stepping off the green westwards, a marvellous view presents itself: you are on the end of a striding-edge with an almost sheer drop to the right. Down below is a great hidden valley containing *another whole golf links!* You can have no real conception of the small grandeur till you see it. I must emphasise again the *scale* of these ranges of dunes at Portrush—they are Alpine, Apennine, *foreign*. On this fourteenth tee you are standing upon the point of a V, the right-hand arm goes out guarding the coast; in the middle is the valley, and along the left-hand arm, exactly along its *edge*, is this next hole called with tactful understatement 'Calamity Corner'. It is two hundred yards long and there is no room on the left, which is the wild rushy crest of the ridge. A yard to the right and over you go, sheer down this grass cliff whose gradient must be 1 in 1. It is a romantic chasm, an opening of hell, and to escape from it it is necessary to defy gravity, taking your stance almost like a fly on a wall, and striking a rocket-shot up to the zenith of heaven. You can do this any number of times, for I almost dare not say that the lies on this cold hill's side are not, well, exactly favourable to pyrotechnics. Having encompassed this, we play 'Purgatory' which swoops all the way down the landward side of the ridge in a superb toboggan-run to a green at the bottom—another less-feathery bed. The drive at Purgatory is very thrilling, for the tee is just below the horizon-line and you hit madly into thin air—the ball disappears away down almost like dropping a penny into a well. These three holes 13–14–15 should be done in 4, 3, 4!

The 16th is a very fine long two-shot hole, notable for its tee, which is sixty-five yards long. If one is at the back end, it is rather like the runway of an aircraft-carrier. The hole doglegs to the right and your second must carry some diagonal traps before reaching safety. There's a grass gully to the left of the green, and a bank on the right.

The finish is a little disappointing after the adventures and explorations we have had. True, the drive at the 17th ought to disappear over the end of a long hog's-back or steer exactly between it and a great rearing sandhill whose near face is all bunker; but once through this Scylla and Charybdis you come out on to the flat, on to almost a field, an inland-looking stretch of harmless grass-and-bunkers. These last two holes are both long, and two steady 5's are not beyond the bounds of probability. But left with two 5's to *win*, I can see them dwindling into 6's without one's quite knowing how or why. If Portrush can be said to have any bad holes, I must record that I think the 18th is at any rate not a good hole! It is, so to speak, comforting, but *officially* comforting and without affection, like an income-tax rebate or the kiss of a strange aunt.

The Championship Committee was right as can be to choose this links for the first Open Championship to be held beyond the sea.   PATRIC DICKINSON, *A Round of Golf Courses*, 1951

❖

PRESTWICK

On your left: mounds of heather and whin. Directly in front: waste. Sheer waste. Small and large clumps of it, sheltered by thin layers of fog. And the caddy hands you a driver. The fairway, presuming one is actually there, can't be more than 20 yards wide, but the caddy hands you a driver.

'*Where* is it?' I asked.

'Straightaway, sir', said Charles, who was distinguished from my caddy at Turnberry by two things. Charles wore a muffler and had his own cigarettes. 'It's just there', he said. 'Just to the left of the cemetery.'

It is asking a lot, I know, to expect anyone to believe that you can bust a drive about 250 yards on a 339-yard hole, have a good lie in the fairway, and still not be able to see a green anywhere, but this is Prestwick.

The green was there, all right, as are all of the greens at Prestwick, but you never see them until you are on them, which is usually eight or ten strokes after leaving the tee. They sit behind little hills, or the terrain simply sinks ten or 15 feet straight down to a mowed surface, or they are snuggled behind tall wood fences over which you have nothing to aim at but a distant church steeple.

You would like to gather up several holes from Prestwick and mail them to your top ten enemies. I guess my all-time favourite love-hate golf hole must be the third hole on this course. Like most of the holes at Prestwick, it is unchanged from the day in 1860 when Willie Park, Sr, shot 174 to become the first Open champion. Quite a score, I have since decided.

First of all, without a caddy, it would take you a week and a half to find the third tee. It is a little patch of ground roughly three yards wide perched atop a stream, a burn, rather, with the cemetery to your back and nothing up ahead except fine mist. Well, dimly in the distance, you can see a rising dune with a fence crawling across it—'the Sleepers', the caddy says. But nothing more. Nothing.

'I'll be frank, Charles', I said. 'I have no idea which way to go, or what with.'

'Have a go with the spoon, sir', he said.

'The spoon?' I shrieked. 'Where the hell am I going with a spoon?'

'A spoon'll get you across the burn, sir, but it'll na get you to the Sleepers', he said.

'Hold it', I said. 'Just wait a minute.' My body was sort of slumped over, and I was holding the bridge of my nose with my thumb and forefinger. 'These, uh, Sleepers. They're out there somewhere?'

'Aye, the Sleepers', he said.

'And, uh, they just kind of hang around, right?'

'Aye', he said. 'The Sleepers have took many a golfer.'

Somehow, I kept the 3-wood in play and when I reached the shot, Charles casually handed me the 4-wood. I took the club and addressed the ball, hoping to hit quickly and get on past the Sleepers, wherever they were. But Charles stopped me.

'Not that way, sir', he said.

'This is the way I was headed when we left the tee', I said.

'We go a bit right here, sir', he said. 'The Sleepers is there just below the old fence. You want to go over the Sleepers and over the fence as well, but na too far right because of the burn. Just a nice stroke, sir, with the 4-wood.'

Happily, I got the shot up and in the general direction Charles ordered, and walking towards the flight of the ball, I finally came to the Sleepers. They were a series of bunkers about as deep as the Grand Canyon. A driver off the tee would have found them, and so would any kind of second shot that didn't get up high enough to clear the fence on the dune. A worn path led through the Sleepers, and then some ancient wooden steps led up the hill and around the fence to what was supposed to be more fairway on the other side.

It wasn't a fairway at all. It was a group of grass moguls going off into infinity. It looked like a carefully arranged assortment of tiny green astrodomes. When Charles handed me the pitching wedge, I almost hit him with it because there was no green in sight.

I got the wedge onto the green that was, sure enough, nestled down in one of those dips, and two-putted for a five that I figured wasn't a par just because the hole was 505 yards long. Charles said I had played the hole perfectly, thanks to him, and that I could play it a thousand times and probably never play it as well.

I said, 'Charles, do you know what this hole would be called in America?'.

'Sir?' he said.

'This is one of those holes where your suitcase flies open and you don't know what's liable to come out', I said.

'Aye, 'tis that', he said.

'One bad shot and you're S.O.L. on this mother', I said.

'Sir?' said Charles.

'Shit out of luck', I said.

'Aye', said Charles. 'At Prestwick, we call it the Sleepers.'

DAN JENKINS, The Dogged Victims of Inexorable Fate, 1970

GOLF IN FICTION

❖

Sir Walter Simpson's maxim that it is impossible to write a good novel
about golf still seems to hold true today, but the game has been a rich source
of material for the short story, and perhaps even more for the episode in a
longer piece on a wider theme. Given the stresses imposed by the game and
the fervour with which it is pursued by so many, this is hardly surprising.
Even the most committed of us must admit that the intensity of the addict
does have its humorous side; the pressures of a tight match do reveal the
character of the player in a way that few other sporting experiences can; and
the social and natural backgrounds to the game are full of potential. But
just as golf has served writers well, so they in turn have by their skills and
imagination illuminated its delights and its disappointments in ways which
no reporter can ever hope to do.

A GOLFING NOVEL

I have seen a golfing novel indeed; but it was in manuscript, the
publishers having rejected it. The scene was St Andrews. He was
a soldier, a statesman, an orator, but only a seventh-class golfer.
She, being St Andrews born, naturally preferred a rising player.
Whichever of the two made the best medal score was to have her
hand. The soldier employed a lad to kick his adversary's ball into
bunkers, to tramp it into mud, to lose it, and he won; but the lady
would not give her hand to a score of 130. Six months passed, during which the soldier studied the game morning, noon, and night,
but to little purpose. Next medal day arrived, and he was face to
face with the fact that his golf, unbacked by his statesmanship,
would avail him nothing. He hired and disguised a professional in
his own clothes. The ruse was successful; but, alas! the professional
broke down. The soldier, disguised as a marker, however, cheated,
and brought him in with 83. A three for the long hole roused suspicion, and led to inquiry. He was found out, dismissed from the club,
rejected by the lady (who afterwards made an unhappy marriage

with a left-handed player), and sent back in disgrace to his statesmanship and oratory. It was as good a romance as could be made
on the subject, but very improbable.

WALTER SIMPSON, *The Art of Golf*, 1887

*The eclectic score of even a moderate performer round his own course is
likely to be well below the official record. If only he could just put it all
together in one single glorious round.*

Ever since the historic day when a visiting clergyman accomplished
the feat of pulling a ball from the tenth tee at an angle of two hundred and twenty-five degrees into the river that is the rightful re-
ceptacle for the eighth tee, the Stockbridge golf course has had
seventeen out of eighteen holes that are punctuated with water
hazards. The charming course itself lies in the flat of the sunken
meadows which the Housatonic, in the few thousand years which
are necessary for the proper preparation of a golf course, has oblig-
ingly eaten out of the high accompanying bluffs. The river, which
goes wriggling on its way as though convulsed with merriment, is
garnished with luxurious elms and willows, which occasionally
deflect to the difficult putting greens the random slices of certain
notorious amateurs.

From the spectacular bluffs of the educated village of Stockbridge
nothing can be imagined more charming than the panorama that
the course presents on a busy day. Across the soft green stretches,
diminutive caddies may be seen scampering with long buckling
nets, while from the riverbanks numerous recklessly exposed legs
wave in the air as the more socially presentable portions hang fran-
tically over the swirling current. Occasionally an enthusiastic golfer,
driving from the eighth or ninth tees, may be seen to start immediately in headlong pursuit of a diverted ball, the swing of the club
and the intuitive leap of the legs forward forming so continuous a
movement that the main purpose of the game often becomes
obscured to the mere spectator. Nearer, in the numerous languid
swales that nature has generously provided to protect the interests
of the manufacturers, or in the rippling patches of unmown grass,
that in the later hours will be populated by enthusiastic caddies,
desperate groups linger in botanizing attitudes.

Every morning lawyers who are neglecting their clients, doctors
who have forgotten their patients, businessmen who have sacrificed

[194]

their affairs, even ministers of the gospel who have forsaken their churches gather in the noisy dressing room and listen with servile attention while some unscrubbed boy who goes around under eighty imparts a little of his miraculous knowledge.

Two hours later, for every ten that have gone out so blithely, two return crushed and despondent, denouncing and renouncing the game, once and for all, absolutely and finally, until the afternoon, when they return like thieves in the night and venture out in a desperate hope; two more come stamping back in even more offensive enthusiasm; and the remainder straggle home moody and disillusioned, reviving their sunken spirits by impossible tales of past accomplishments.

There is something about these twilight gatherings that suggests the degeneracy of a rugged race; nor is the contamination of merely local significance. There are those who lie consciously, with a certain frank, commendable, wholehearted plunge into iniquity. Such men return to their worldly callings with intellectual vigor unimpaired and a natural reaction toward the decalogue. Others of more casuistical temperament, unable all at once to throw over the traditions of a New England conscience to the exigencies of the game, do not at once burst into falsehood, but by a confusing process weaken their memories and corrupt their imaginations. They never lie of the events of the day. Rather they return to some jumbled happening of the week before and delude themselves with only a lingering qualm, until from habit they can create what is really a form of paranoia, the delusion of greatness, or the exaggerated ego. Such men, inoculated with self-deception, return to the outer world to deceive others, lower the standards of business morality, contaminate politics, and threaten the vigor of the republic.

R. N. Booverman, the treasurer, and Theobald Pickings, the unenvied secretary of an unenvied board, arrived at the first tee at precisely ten o'clock on a certain favorable morning in early August to begin the thirty-six holes which six times a week, six months of the year, they played together as sympathetic and well-matched adversaries. Their intimacy had arisen primarily from the fact that Pickings was the only man willing to listen to Booverman's restless dissertations on the malignant fates which seemed to pursue him even to the neglect of their international duties, while Booverman, in fair exchange, suffered Pickings to enlarge *ad libitum* on his theory of the rolling versus the flat putting greens.

[195]

Pickings was one of those correctly fashioned and punctilious golfers whose stance was modeled on classic lines, whose drive, though it averaged only twenty-five yards over the hundred, was always a well-oiled and graceful exhibition of the Royal St Andrew's swing, the left sole thrown up, the eyeballs bulging with the last muscular tension, the club carried back until the whole body was contorted into the first position of the traditional hoop snake preparing to descend a hill. He used the interlocking grip, carried a bag with a spoon driver, an aluminum cleek, had three abnormal putters, and wore one chamois glove with air holes on the back. He never accomplished the course in less than eighty-five and never exceeded ninety-four, but, having aimed to set a correct example rather than to strive vulgarly for professional records, was always in a state of offensive optimism due to a complete sartorial satisfaction.

Booverman, on the contrary, had been hailed in his first years as a coming champion. With three holes eliminated, he could turn in a card distinguished for its fours and threes; but unfortunately these sad lapses inevitably occurred. As Booverman himself admitted, his appearance on the golf links was the signal for the capricious imps of chance who stir up politicians to indiscreet truths and keep the Balkan pot of discord bubbling, to forsake immediately these prime duties, and enjoy a little relaxation at his expense.

After some delay they drive off.

'Fine shot, Mr Booverman,' said Frank, the professional, nodding his head, 'free and easy, plenty of follow-through.'

'You're on your drive today,' said Pickings, cheerfully.

'Sure! When I get a good drive off the first tee,' said Booverman, discouraged, 'I mess up all the rest. You'll see.'

'Oh, come now,' said Pickings, as a matter of form. He played his shot, which came methodically to the edge of the green.

Booverman took his mashy for the short running-up stroke to the pin, which seemed so near.

'I suppose I've tried this shot a thousand times,' he said savagely. 'Anyone else would get a three once in five times—anyone but Jonah's favorite brother.'

He swung carelessly, and watched with a tolerant interest the white ball roll on to the green straight for the flag. All at once

Wessels and Pollock, who were ahead, sprang into the air and began agitating their hats.

'By George! It's in!' said Pickings. 'You've run it down. First hole in two! Well, what do you think of that?'

Booverman, unconvinced, approached the hole with suspicion, gingerly removing the pin. At the bottom, sure enough, lay his ball for a phenomenal two.

'That's the first bit of luck that has ever happened to me,' he said furiously, 'absolutely the first time in my whole career.'

'I say, old man,' said Pickings in remonstrance, 'you're not angry about it, are you?'

'Well, I don't know whether I am or not,' said Booverman obstinately. In fact, he felt rather defrauded. The integrity of his record was attacked. 'See here, I play thirty-six holes a day, two hundred and sixteen a week, a thousand a month, six thousand a year; ten years, sixty thousand holes; and this is the first time a bit of luck has ever happened to me—once in sixty thousand times.'

Pickings drew out a handkerchief and wiped his forehead.

'It may come all at once,' he said faintly.

This mild hope only infuriated Booverman. He had already teed his ball for the second hole, which was poised on a rolling hill one hundred and thirty-five yards away. It is considered rather easy as golf holes go. The only dangers are a matted wilderness of long grass in front of the tee, the certainty of landing out of bounds on the slightest slice, and of rolling down hill into a soggy substance on a pull. Also there is a tree to be hit and a sand pit to be sampled.

'Now watch my little friend the apple tree,' said Booverman. 'I'm going to play for it, because, if I slice, I lose my ball, and that knocks my whole game higher than a kite.' He added between his teeth: 'All I ask is to get around to the eighth hole before I lose my ball. I know I'll lose it there.'

Due to the fact that his two on the first brought him not the slightest thrill of nervous joy, he made a perfect shot, the ball carrying the green straight and true.

'This is your day, all right,' said Pickings, stepping to the tee.

'Oh, there's never been anything the matter with my irons,' said Booverman darkly. 'Just wait till we strike the fourth and fifth holes.'

When they climbed the hill, Booverman's ball lay within three feet of the cup, which he easily putted out.

'Two down,' said Pickings inaudibly. 'By George! What a glorious start!'

'Once in sixty thousand times,' said Booverman to himself. The third hole lay two hundred and five yards below, backed by the road and trapped by ditches, where at that moment Pollock, true to his traditions as a war correspondent, was laboring in the trenches, to the unrestrained delight of Wessels, who had passed beyond.

'Theobald,' said Booverman, selecting his cleek and speaking with inspired conviction, 'I will tell you exactly what is going to happen. I will smite this little homeopathic pill, and it will land just where I want it. I will probably put out for another two. Three holes in twos would probably excite any other human being on the face of this globe. It doesn't excite me. I know too well what will follow on the fourth or fifth watch.'

'Straight to the pin,' said Pickings in a loud whisper. 'You've got a dead line on every shot today. Marvelous! When you get one of your streaks, there's certainly no use in my playing.'

'Streak's the word,' said Booverman, with a short, barking laugh. 'Thank heaven, though, Pickings, I know it! Five years ago I'd have been shaking like a leaf. Now it only disgusts me. I've been fooled too often; I don't bite again.'

In this same profoundly melancholy mood he approached his ball, which lay on the green, hole high, and put down a difficult putt, a good three yards for his third two.

Pickings, despite all his classic conservatism, was so overcome with excitement that he twice putted over the hole for a shameful five.

Booverman's face as he walked to the fourth tee was as joyless as a London fog. He placed his ball carelessly, selected his driver, and turned on the fidgety Pickings with the gloomy solemnity of a father about to indulge in corporal punishment.

'Once in sixty thousand times, Picky. Do you realize what a start like this—three twos—would mean to a professional like Frank or even an amateur that hadn't offended every busy little fate and fury in the whole hoodooing business? Why, the blooming record would be knocked into the middle of next week.'

'You'll do it,' said Pickings in a loud whisper. 'Play carefully.'

Finally they reach the twelfth . . .

The twelfth hole is another dip into the long grass that might serve as an elephant's bed, and then across the Housatonic River, a carry of one hundred and twenty yards to the green at the foot of an intruding tree.

'Oh, I suppose I'll make another three here, too,' said Booverman moodily. 'That'll only make it worse.'

He drove with his midiron high in the air and full on the flag.

'I'll play my putt carefully for a three,' he said, nodding his head. Instead, it ran straight and down for a two.

He walked silently to the dreaded thirteenth tee, which, with the returning fourteenth, forms the malignant Scylla and Charybdis of the course. There is nothing to describe the thirteenth hole. It is not really a golf hole; it is a long, narrow breathing spot, squeezed by the railroad tracks on one side and by the river on the other. Resolute and fearless golfers often cut them out entirely, nor are ashamed to acknowledge their terror. As you stand at the thirteenth tee, everything is blurred to the eye. Nearby are rushes and water, woods to the left and right; the river and the railroad and the dry land a hundred yards away look tiny and distant, like a rock amid floods.

A long drive that varies a degree is doomed to go out of bounds or to take the penalty of the river.

'Don't risk it. Take an iron—play it carefully,' said Pickings in a voice that sounded to his own ears unrecognizable.

Booverman followed his advice and landed by the fence to the left, almost off the fair. A midiron for his second put him in position for another four, and again brought his score to even threes.

When the daring golfer has passed quaking up the narrow way and still survives, he immediately falls a victim to the fourteenth, which is a bend hole, with all the agonies of the preceding thirteenth, augmented by a second shot over a long, mushy pond. If you play a careful iron to keep from the railroad, now on the right, or to dodge the river on your left, you are forced to approach the edge of the swamp with a cautious fifty-yard-running-up stroke before facing the terrors of the carry. A drive with a wooden club is almost sure to carry into the swamp, and only a careful cleek shot is safe.

'I wish I were playing this for the first time,' said Booverman, blackly. 'I wish I could forget—rid myself of memories. I have seen class A amateurs take twelve, and professionals eight. This is the end of all things, Picky, the saddest spot on earth. I won't waste time. Here goes.'

To Pickings's horror, the drive began slowly to slice out of bounds, toward the railroad tracks.

'I knew it,' said Booverman calmly, 'and the next will go there, too; then I'll put one in the river, two in the swamp, slice into—'

All at once he stopped, thunderstruck. The ball, hitting tie or rail, bounded high in the air, forward, back upon the course, lying in perfect position.

Pickings said something in a purely reverent spirit.

'Twice in sixty thousand times,' said Booverman, unrelenting. 'That only evens up the sixth hole. Twice in sixty thousand times!'

From where the ball lay an easy brassy brought it near enough to the green to negotiate another four. Pickings, trembling like a toy dog in zero weather, reached the green in ten strokes, and took three more putts.

The fifteenth, a short pitch over the river, eighty yards to a slanting green entirely surrounded by more long grass, which gave it the appearance of a chin spot on a full face of whiskers, was Booverman's favorite hole. While Pickings held his eyes to the ground and tried to breathe in regular breaths, Booverman placed his ball, drove with the requisite backspin, and landed dead to the hole. Another two resulted.

'Even threes—fifteen holes in even threes,' said Pickings to himself, his head beginning to throb. He wanted to sit down and take his temples in his hands, but for the sake of history he struggled on.

'Damn it!' said Booverman all at once.

'What's the matter?' said Pickings, observing his face black with fury.

'Do you realize, Pickings, what it means to me to have lost those two strokes on the fourth and sixth greens, and through no fault of mine, either? Even threes for the whole course—that's what I could do if I had those two strokes—the greatest thing that's ever been seen on a golf course. It may be a hundred years before any human being on the face of this earth will get such a chance. And to think I might have done it with a little luck!'

Pickings felt his heart begin to pump, but he was able to say with some degree of calm:

'You may get a three here.'

'Never. Four, three and four is what I'll end.'

'Well, good heavens! What do you want?'

'There's no joy in it, though,' said Booverman gloomily. 'If I had those two strokes back, I'd go down in history, I'd be immortal. And you, too, Picky, because you went around with me. The fourth hole was bad enough, but the sixth was heart-breaking.'

His drive cleared another swamp and rolled well down the farther plateau. A long cleek laid his ball off the green, a good approach stopped a little short of the hole, and the putt went down.

'Well, that ends it,' said Booverman, gloomily. 'I've got to make a two or three to do it. The two is quite possible; the three absurd.'

The seventeenth hole returns to the swamp that enlivens the sixth. It is a full cleek, with about six mental hazards distributed in Indian ambush, and in five of them a ball may lie until the day of judgment before rising again.

Pickings turned his back, unable to endure the agony of watching. The click of the club was sharp and true. He turned to see the ball in full flight arrive unerringly hole high on the green.

'A chance for a two,' he said under his breath. He sent two balls into the lost land to the left and one into the rough to the right.

'Never mind me,' he said, slashing away in reckless fashion.

Booverman, with a little care, studied the ten-foot route to the hole and putted down.

'Even threes!' said Pickings, leaning against a tree.

'Blast that sixth hole!' said Booverman, exploding. 'Think of what it might be, Picky—what it ought to be!'

Pickings retired hurriedly before the shaking approach of Booverman's frantic club. Incapable of speech, he waved him feebly to drive. He began incredulously to count up again, as though doubting his senses.

'One under three, even threes, one over, even, one under—'

'What the deuce are you doing?' said Booverman angrily. 'Trying to throw me off?'

'I didn't say anything,' said Pickings.

'You didn't—muttering to yourself.'

'I must make him angry, keep his mind off the score,' said Pickings feebly to himself. He added aloud, 'Stop kicking about your old sixth hole! You've had the darndest luck I ever saw, and yet you grumble.'

Booverman swore under his breath, hastily approached his ball, drove perfectly, and turned in a rage.

'Luck?' he cried furiously. 'Pickings, I've a mind to wring your neck. Every shot I've played has been dead on the pin, now, hasn't it?'

'How about the ninth hole—hitting a tree?'

'Whose fault was that? You had no right to tell me my score, and, besides, I only got an ordinary four there, anyway.'

'How about the railroad track?'

'One shot out of bounds. Yes, I'll admit that. That evens up for the fourth.'

'How about your first hole in two?'

'Perfectly played; no fluke about it at all—once in sixty thousand times. Well, any more sneers? Anything else to criticize?'

'Let it go at that.'

Booverman, in this heckled mood, turned irritably to his ball, played a long midiron, just cleared the crescent bank of the last swale, and ran up on the green.

'Damn that sixth hole!' said Booverman, flinging down his club and glaring at Pickings. 'One stroke back, and I could have done it.'

Pickings tried to address his ball, but the moment he swung his club his legs began to tremble. He shook his head, took a long breath, and picked up his ball.

They approached the green on a drunken run in the wild hope that a short putt was possible. Unfortunately the ball lay thirty feet away, and the path to the hole was bumpy and riddled with worm casts. Still, there was a chance, desperate as it was.

Pickings let his bag slip to the ground and sat down, covering his eyes while Booverman with his putter tried to brush away the ridges.

'Stand up!'

Pickings rose convulsively.

'For heaven's sake, Picky, stand up! Try to be a man!' said Booverman hoarsely. 'Do you think I've any nerve when I see you with chills and fever? Brace up!'

'All right.'

Booverman sighted the hole, and then took his stance; but the cleek in his hand shook like an aspen. He straightened up and walked away.

'Picky,' he said, mopping his face, 'I can't do it. I can't putt it.'

'You must.'

'I've got buck fever. I'll never be able to putt it—never.'

At the last, no longer calmed by an invincible pessimism, Booverman had gone to pieces. He stood shaking from head to foot.

'Look at that,' he said, extending a fluttering hand. 'I can't do it; I can never do it.'

'Old fellow, you must,' said Pickings. 'You've got to. Bring yourself together. Here!'

He slapped him on the back, pinched his arms, and chafed his fingers. Then he led him back to the ball, braced him into position, and put the putter in his hands.

'Buck fever,' said Booverman in a whisper. 'Can't see a thing.'

Pickings, holding the flag in the cup, said savagely:

'Shoot!'

The ball advanced in a zigzag path, running from worm cast to worm cast, wobbling and rocking, and at the last, as though preordained, fell plump into the cup!

At the same moment, Pickings and Booverman, as though carried off by the same cannonball, flattened on the green.

Five minutes later, wild-eyed and hilarious, they descended on the clubhouse with the miraculous news. For an hour the assembled golfers roared with laughter as the two stormed, expostulated, and swore to the truth of the tale.

They journeyed from house to house in a vain attempt to find some convert to their claim. For a day they passed as consummate comedians, and the more they yielded to their rage, the more consummate was their art declared. Then a change took place. From laughing the educated town of Stockbridge turned to resentment, then to irritation, and finally to suspicion. Booverman and Pickings began to lose caste, to be regarded as unbalanced, if not positively dangerous. Unknown to them, a committee carefully examined the books of the club. At the next election another treasurer and another secretary were elected.

Since then, month in and month out, day after day, in patient hope, the two discredited members of the educated community of

Stockbridge may be seen, *accompanied by caddies*, toiling around the links in a desperate belief that the miracle that would restore them to standing may be repeated. Each time as they arrive nervously at the first tee and prepare to swing, something between a chuckle and a grin runs through the assemblage, while the left eyes contract waggishly, and a murmuring may be heard:
'Even threes.'

The Stockbridge golf links is a course of ravishing beauty and the Housatonic River, as has been said, goes wriggling around it as though convulsed with merriment.   OWEN JOHNSON, *Even Threes*, 1912

In the following passage by Ford Madox Ford, Chrissie Tietjens and Macmaster, his friend and colleague at the Imperial Department of Statistics, have come down to Rye to play. They meet General Lord Edward Campion, his son-in-law Paul Sandbach, who is the Conservative MP for the division, and, in the clubhouse, the Rt. Hon. Stephen Waterhouse, a Liberal Cabinet Minister who is a particular target for the suffragettes. Also in the club are two 'city men' whose coarse manners and conversation outrage the other members.

The General shook his head:
'You brilliant fellows!' he said. 'The country, or the army, or anything, could not be run by you. It takes stupid fools like me and Sandbach, along with sound moderate heads like our friend here.'
He indicated Macmaster and, rising, went on: 'Come along. You're playing me, Macmaster. They say you're hot stuff. Chrissie's no good. He can take Sandbach on.'
He walked off with Macmaster towards the dressing-room.
Sandbach, wriggling awkwardly out of his chair, shouted:
'Save the country. . . . Damn it. . . .' He stood on his feet. 'I and Campion . . . Look at what the country's come to. . . . What with swine like these two in our club houses! And policemen to go round the links with Ministers to protect them from the wild women. . . . By God! I'd like to have the flaying of the skin off some of their backs I would. My God I would.'
He added:
'That fellow Waterslops is a bit of a sportsman. I haven't been able to tell you about our bet, you've been making such a noise. . . . Is your friend really plus one at North Berwick? What are you like?'

'Macmaster is a good plus two anywhere when he's in practice.'
Sandbach said:
'Good Lord. . . . A stout fellow. . . .'
'As for me,' Tietjens said, 'I loathe the beastly game.'
'So do I,' Sandbach answered. 'We'll just lollop along behind
them.'

They came out into the bright open where all the distances under
the tall sky showed with distinct prismatic outlines. They made
a little group of seven—for Tietjens would not have a caddy—
waiting on the flat, first teeing ground.

Mr Sandbach hobbled from one to the other explaining the terms
of his wager with Mr Waterhouse. Mr Waterhouse had backed one
of the young men playing with him to drive into and hit twice in
the eighteen holes the two city men who would be playing ahead
of them. As the Minister had taken rather short odds, Mr Sandbach
considered him a good sport.

A long way down the first hole Mr Waterhouse and his two companions were approaching the first green. They had high sandhills
to the right and, to their left, a road that was fringed with rushes
and a narrow dyke. Ahead of the Cabinet Minister the two city
men and their two caddies stood on the edge of the dyke or poked
downwards into the rushes. Two girls appeared and disappeared
on the tops of the sandhills. The policeman was strolling along the
road, level with Mr Waterhouse. The General said:
'I think we could go now.'
Sandbach said:
'Waterslops will get a hit at them from the next tee. They're in
the dyke.'

The General drove a straight, goodish ball. Just as Macmaster
was in his swing Sandbach shouted:
'By God! He nearly did it. See that fellow jump!'

Macmaster looked round over his shoulder and hissed with vexa-
tion between his teeth:
'Don't you know that you don't shout while a man is driving?
Or haven't you played golf?' He hurried fussily after his ball.

Sandbach said to Tietjens:
'Golly! That chap's got a temper!'
Tietjens said:
'Only over this game. You deserved what you got.'

Sandbach said:

'I did. . . . But I didn't spoil his shot. He's outdriven the General twenty yards.'

Tietjens said:

'It would have been sixty but for you.'

They loitered about on the tee waiting for the others to get their distance. Sandbach said:

'By Jove, your friend is on with his second. . . . You wouldn't believe it of such a *little* beggar!' He added: 'He's not much class, is he?'

Tietjens looked down his nose.

'Oh, about *our* class!' he said. 'He wouldn't take a bet about driving into the couple ahead.'

Sandbach hated Tietjens for being a Tietjens of Groby: Tietjens was enraged by the existence of Sandbach, who was the son of an ennobled mayor of Middlesbrough, seven miles or so from Groby. The feuds between the Cleveland landowners and the Cleveland plutocrats are very bitter. Sandbach said:

'Ah, I suppose he gets you out of scrapes with girls and the Treasury, and you take him about in return. It's a practical combination.'

'Like Pottle Mills and Stanton,' Tietjens said. The financial operations connected with the amalgamating of these two steelworks had earned Sandbach's father a good deal of odium in the Cleveland district. . . . Sandbach said:

'Look here, Tietjens. . . .' But he changed his mind and said:

'We'd better go now.' He drove off with an awkward action but not without skill. He certainly outplayed Tietjens.

Playing very slowly, for both were desultory and Sandbach very lame, they lost sight of the others behind some coastguard cottages and dunes before they had left the third tee. Because of his game leg Sandbach sliced a good deal. On this occasion he sliced right into the gardens of the cottages and went with his boy to look for his ball among potato-haulms, beyond a low wall. Tietjens patted his own ball lazily up the fairway and, dragging his bag behind him by the strap, he sauntered on.

Although Tietjens hated golf as he hated any occupation that was of a competitive nature, he could engross himself in the mathematics of trajectories when he accompanied Macmaster in one of his expeditions for practice. He accompanied Macmaster because he liked there to be one pursuit at which his friend undisputably excelled himself, for it was a bore always brow-beating the fellow. But he stipulated that they should visit three different and, if possible, unknown courses every week-end when they golfed. He interested himself then in the way the courses were laid out, acquiring thus an extraordinary connoisseurship in golf architecture, and he made abstruse calculations as to the flight of balls off sloped club-faces, as to the foot-poundals of energy exercised by one muscle or the other, and as to theories of spin. As often as not he palmed Macmaster off as a fair, average player on some other unfortunate fair, average stranger. Then he passed the afternoon in the club-house studying the pedigrees and forms of racehorses, for every club-house contained a copy of Ruff's Guide. In the spring he would hunt for and examine the nests of soft-billed birds, for he was interested in the domestic affairs of the cuckoo, though he hated natural history and field botany.

On this occasion he had just examined some notes of other mashie shots, had put the notebook back in his pocket, and had addressed his ball with a niblick that had an unusually roughened face and a head like a hatchet. Meticulously, when he had taken his grip he removed his little and third fingers from the leather of the shaft. He was thanking heaven that Sandbach seemed to be accounted for for ten minutes at least, for Sandbach was miserly over lost balls and, very slowly, he was raising his mashie to half cock for a sighting shot.

He was aware that someone, breathing a little heavily from small lungs, was standing close to him and watching him: he could indeed, beneath his cap-rim, perceive the tips of a pair of boy's white sand-shoes. It in no way perturbed him to be watched, since he was avid of no personal glory when making his shots. A voice said:

'I say . . .' He continued to look at his ball.

'Sorry to spoil your shot,' the voice said. 'But . . .'

Tietjens dropped his club altogether and straightened his back. A fair young woman with a fixed scowl was looking at him intently. She had a short skirt and was panting a little.

'I say,' she said, 'go and see they don't hurt Gertie. I've lost her . . .' She pointed back to the sandhills. 'There looked to be some beasts among them.'

She seemed a perfectly negligible girl except for the frown: her eyes blue, her hair no doubt fair under a white canvas hat. She had a striped cotton blouse, but her fawn tweed skirt was well hung.

Tietjens said:

'You've been demonstrating.'

She said:

'Of course we have, and of course you object on principle. But you won't let a girl be man-handled. Don't wait to tell me, I know it. . . . '

Noises existed. Sandbach, from beyond the low garden wall fifty yards away, was yelping, just like a dog: 'Hi! Hi! Hi! Hi!' and gesticulating. His little caddy, entangled in his golfbag, was trying to scramble over the wall. On top of a high sandhill stood the policeman: he waved his arms like a windmill and shouted. Beside him and behind, slowly rising, were the heads of the General, Macmaster and their two boys. Farther along, in completion, were appearing the figures of Mr Waterhouse, his two companions and *their* three boys. The Minister was waving his driver and shouting. They all shouted.

'A regular rat-hunt,' the girl said; she was counting. 'Eleven and two more caddies!' She exhibited satisfaction. 'I headed them all off except two beasts. They couldn't run. But neither can Gertie. . . . '

She said urgently:

'Come along! You aren't going to leave Gertie to those beasts! They're drunk. . . . '

Tietjens said:

'Cut away then. I'll look after Gertie.' He picked up his bag.

'No, I'll come with you,' the girl said.

Tietjens answered: 'Oh, you don't want to go to gaol. Clear out!'

She said:

'Nonsense. I've put up with worse than that. Nine months as a slavey. . . . Come *along!*'

Tietjens started to run—rather like a rhinoceros seeing purple. He had been violently spurred, for he had been pierced by a shrill, faint scream. The girl ran beside him.

'You . . . can . . . run!' she panted, 'put on a spurt.'

Screams protesting against physical violence were at that date rare things in England. Tietjens had never heard the like. It upset him frightfully, though he was aware only of an expanse of open country. The policeman, whose buttons made him noteworthy, was descending his conical sandhill, diagonally, with caution. There is something grotesque about a town policeman, silvered helmet and

all, in the open country. It was so clear and still in the air; Tietjens
felt as if he were in a light museum looking at specimens. . . .

A little young woman, engrossed, like a hunted rat, came round
the corner of a green mound. 'This is an assaulted female!' the
mind of Tietjens said to him. She had a black skirt covered with
sand, for she had just rolled down the sandhill; she had a striped
grey and black silk blouse, one shoulder torn completely off, so
that a white camisole showed. Over the shoulder of the sandhill
came the two city men, flushed with triumph and panting; their
red knitted waistcoats moved like bellows. The black-haired one,
his eyes lurid and obscene, brandished aloft a fragment of black
and grey stuff. He shouted hilariously:

'Strip the bitch naked! . . . Ugh . . . Strip the bitch stark naked!'
and jumped down the little hill. He cannoned into Tietjens, who
roared at the top of his voice:

'You infernal swine. I'll knock your head off if you move!'

Behind Tietjens' back the girl said:

'Come along, Gertie. . . . It's only to there . . . '

A voice panted in answer:

'I . . . can't. . . . My heart . . . '

Tietjens kept his eye upon the city man. His jaw had fallen down,
his eyes stared! It was as if the bottom of his assured world, where
all men desire in their hearts to bash women, had fallen out. He
panted:

'Ergle! Ergle!'

Another scream, a little farther than the last voices from behind
his back, caused in Tietjens a feeling of intense weariness. What
did beastly women want to scream for? He swung round, bag and
all. The policeman, his face scarlet like a lobster just boiled, was
lumbering unenthusiastically towards the two girls who were trotting towards the dyke. One of his hands, scarlet also, was extended.
He was not a yard from Tietjens.

Tietjens was exhausted, beyond thinking or shouting. He slipped
his clubs off his shoulder and, as if he were pitching his kit-bag
into a luggage van, threw the whole lot between the policeman's
running legs. The man, who had no impetus to speak of, pitched
forward on to his hands and knees. His helmet over his eyes, he
seemed to reflect for a moment; then he removed his helmet and
with great deliberation rolled round and sat on the turf. His face
was completely without emotion, long, sandy-moustached and

rather shrewd. He mopped his brow with a carmine handkerchief that had white spots.

Tietjens walked up to him.

'Clumsy of me!' he said. 'I hope you're not hurt.' He drew from his breast pocket a curved silver flask. The policeman said nothing. His world, too, contained uncertainties, and he was profoundly glad to be able to sit still without discredit. He muttered:

'Shaken. A bit! Anybody would be!'

That let him out and he fell to examining with attention the bayonet catch of the flask top. Tietjens opened it for him. The two girls, advancing at a fatigued trot, were near the dyke side. The fair girl, as they trotted, was trying to adjust her companion's hat; attached by pins to the back of her hair it flapped on her shoulder.

All the rest of the posse were advancing at a very slow walk, in a converging semi-circle. Two little caddies were running, but Tietjens saw them check, hesitate and stop. And there floated to Tietjens' ears the words:

'Stop, you little devils. She'll knock your heads off.'

The Rt. Hon. Mr Waterhouse must have found an admirable voice trainer somewhere. The drab girl was balancing tremulously over a plank on the dyke; the other took it at a jump; up in the air—down on her feet; perfectly business-like. And, as soon as the other girl was off the plank, she was down on her knees before it, pulling it towards her, the other girl trotting away over the vast marsh field.

The girl dropped the plank on the grass. Then she looked up and faced the men and boys who stood in a row on the road. She called in a shrill, high voice, like a young cockerel's:

'Seventeen to two! The usual male odds! You'll *have* to go round by Camber railway bridge, and we'll be in Folkestone by then. We've got bicycles!' She was half going when she checked and, searching out Tietjens to address, exclaimed: 'I'm sorry I said that. Because some of you didn't want to catch us. But some of you *did*. And you *were* seventeen to two.' She addressed Mr Waterhouse:

'Why *don't* you give women the vote?' she said. 'You'll find it will interfere a good deal with your indispensable golf if you don't. Then what becomes of the nation's health?'

Mr Waterhouse said:

'If you'll come and discuss it quietly . . .'

She said:

'Oh, tell that to the marines,' and turned away, the men in a row watching her figure disappear into the distance of the flat land. Not one of them was inclined to risk that jump: there was nine foot of mud in the bottom of the dyke. It was quite true that, the plank being removed, to go after the women they would have had to go several miles round. It had been a well-thought-out raid. Mr Waterhouse said that girl was a ripping girl: the others found her just ordinary. Mr Sandbach, who had only lately ceased to shout: 'Hi!' wanted to know what they were going to do about catching the women, but Mr Waterhouse said: 'Oh, chuck it, Sandy,' and went off.

Mr Sandbach refused to continue his match with Tietjens. He said that Tietjens was the sort of fellow who was the ruin of England. He said he had a good mind to issue a warrant for the arrest of Tietjens—for obstructing the course of justice. Tietjens pointed out that Sandbach wasn't a borough magistrate and so couldn't. And Sandbach went off, dot and carry one, and began a furious row with the two city men who had retreated to a distance. He said they were the sort of men who were the ruin of England. They bleated like rams. . . .

Tietjens wandered slowly up the course, found his ball, made his shot with care and found that the ball deviated several feet less to the right of a straight line than he had expected. He tried the shot again, obtained the same result and tabulated his observations in his notebook. He sauntered slowly back towards the club-house. He was content.

He felt himself to be content for the first time in four months. His pulse beat calmly; the heat of the sun all over him appeared to be a beneficent flood. On the flanks of the older and larger sandhills he observed the minute herbage, mixed with little purple aromatic plants. To these the constant nibbling of sheep had imparted a protective tininess. He wandered, content, round the sandhills to the small, silted harbour mouth.

FORD MADOX FORD, *Some do not*, 1924

P. G. WODEHOUSE

I had often wondered what P. G. Wodehouse's own golf was like until I recently came upon a reference to it in E. Phillips Oppenheim's autobiography:

[211]

Plum Wodehouse's golf was, and would be still, I expect, if he had a chance to play, of a curious fashion. He had only one idea in his mind when he took up his stance on the tee, and that idea was length. He was almost inattentive when his caddie pointed out the line he ought to take or the actual whereabouts of the next hole, but he went for the ball with one of the most comprehensive and vigorous swings I have ever seen. I am certain that I saw him hit a ball once at Woking which was the longest shot I have ever seen in my life without any trace of following wind. It was, I believe, on a Sunday morning, from the seventeenth tee.

'You will never see that again!' I remarked, after my first gasp of astonishment, mingled, I am afraid I must confess, with a certain amount of malevolent pleasure as the ball disappeared in the bosom of a huge clump of gorse.

'I wonder how far it was,' was the wistful reply.

Well, the Wodehouses were spending the week-end, and I noticed after we arrived at the club-house on the conclusion of our round a mysterious conversation going on between P. G. and his caddie. Late in the evening, the caddie was ushered into my garden. He produced a ball and handed it over.

'Found it half an hour ago, I did, sir,' he remarked.

'And did you put the stick in?' P. G. asked eagerly.

'Right where the ball lay to an inch, sir.'

'Got the distance?'

'Three hundred and forty-three yards, sir,' the caddie replied promptly.

There was a glow of happiness in P. G.'s expression. He dragged me down to see where the ball had been found and checked the distance going back. Then he filled a pipe and was very happy.

'Beaten my own record by five yards,' he confided with a grin.

'But listen,' I pointed out, 'how many matches do you win?'

'I never win a match,' was the prompt reply. 'I spend my golfing life out of bounds. I never even count my strokes. I know that I can never beat anyone who putts along down the middle. All the same I get more fun out of my golf than any other man I know when I am hitting my drives.'

E. PHILLIPS OPPENHEIM, The Pool of Memory, 1941

However that may be, there is certainly no one who has made better use of the game as a setting for humorous writing. The story from which the following extract is taken is told, as ever, by the Oldest Member, and is an illustration of his maxim that 'to refrain entirely from oaths during a round is almost equivalent to giving away three bisques'.

Chester Meredith has fallen in love with Felicia Blakeney, a fellow member of his golf club. In order to impress her he not only pretends to be a friend of her brother (whom she secretly loathes), but also, during play, represses his normal language and limits himself to an occasional feeble giggle. She interprets this behaviour as evidence of unmanly stupidity and rejects his proposal of marriage.

I saw him the day after he had been handed the mitten, and was struck by the look of grim determination in his face. Deeply wounded though he was, I could see that he was the master of his fate and the captain of his soul.

'I am sorry, my boy,' I said, sympathetically, when he had told me the painful news.

'It can't be helped,' he replied, bravely.

'Her decision was final?'

'Quite.'

'You do not contemplate having another pop at her?'

'No good. I know when I'm licked.'

I patted him on the shoulder and said the only thing it seemed possible to say.

'After all, there is always golf.'

He nodded.

'Yes. My game needs a lot of tuning up. Now is the time to do it. From now on I go at this pastime seriously. I make it my life-work. Who knows?' he murmured, with a sudden gleam in his eyes. 'The Amateur Championship——'

'The Open!' I cried, falling gladly into his mood.

'The American Amateur,' said Chester, flushing.

'The American Open,' I chorused.

'No one has ever copped all four.'

'No one.'

'Watch me!' said Chester Meredith, simply.

It was about two weeks after this that I happened to look in on Chester at his house one morning. I found him about to start for the links. As he had foreshadowed in the conversation which I have just related, he now spent most of the daylight hours on the course. In these two weeks he had gone about his task of achieving perfection with a furious energy which made him the talk of the club. Always one of the best players in the place, he had developed an astounding brilliance. Men who had played him level were now obliged to receive two and even three strokes. The pro. himself, conceding one, had only succeeded in halving their match. The struggle for the President's Cup came round once more, and Chester won it for the second time with ridiculous ease.

When I arrived, he was practising chip-shots in his sitting-room. I noticed that he seemed to be labouring under some strong emotion, and his first words gave me the clue.

'She's going away tomorrow,' he said, abruptly, lofting a ball over the whatnot on to the Chesterfield.

I was not sure whether I was sorry or relieved. Her absence would leave a terrible blank, of course, but it might be that it would help him to get over his infatuation.

'Ah!' I said, non-committally.

Chester addressed his ball with a well-assumed phlegm, but I could see by the way his ears wiggled that he was feeling deeply. I was not surprised when he topped his shot into the coal-scuttle.

'She has promised to play a last round with me this morning,' he said.

Again I was doubtful what view to take. It was a pretty, poetic idea, not unlike Browning's 'Last Ride Together', but I was not sure if it was altogether wise. However, it was none of my business, so I merely patted him on the shoulder and he gathered up his clubs and went off.

Owing to motives of delicacy I had not offered to accompany him on his round, and it was not till later that I learned the actual details of what occurred. At the start, it seems, the spiritual anguish which he was suffering had a depressing effect on his game. He hooked his drive off the first tee and was only enabled to get a five by means of a strong niblick shot out of the rough. At the second, the lake hole, he lost a ball in the water and got another five. It was only at the third that he began to pull himself together.

The test of a great golfer is his ability to recover from a bad start. Chester had this quality to a pre-eminent degree. A lesser man, conscious of being three over bogey for the first two holes, might have looked on his round as ruined. To Chester it simply meant that he had to get a couple of 'birdies' right speedily, and he set about it at once. Always a long driver, he excelled himself at the third. It is, as you know, an uphill hole all the way, but his drive could not have come far short of two hundred and fifty yards. A brassie-shot of equal strength and unerring direction put him on the edge of the green, and he holed out with a long putt two under bogey. He had hoped for a 'birdie' and he had achieved an 'eagle'.

I think that this splendid feat must have softened Felicia's heart, had it not been for the fact that misery had by this time entirely robbed Chester of the ability to smile. Instead, therefore, of behaving in the wholesome, natural way of men who get threes at bogey-five holes, he preserved a drawn, impassive countenance; and as she watched him tee up her ball, stiff, correct, polite, but to all outward appearance absolutely inhuman, the girl found herself stifling that thrill of what for a moment had been almost adoration. It was, she felt, exactly how her brother Crispin would have comported himself if he had done a hole in two under bogey.

And yet she could not altogether check a wistful sigh when, after a couple of fours at the next two holes, he picked up another stroke on the sixth and with an inspired spoon-shot brought his medal-score down to one better than bogey by getting a two at the hundred-and-seventy-yard seventh. But the brief spasm of tenderness passed, and when he finished the first nine with two more fours she refrained from anything warmer than a mere word of stereotyped congratulation.

'One under bogey for the first nine,' she said. 'Splendid!'

'One under bogey!' said Chester, woodenly.

'Out in thirty-four. What is the record for the course?'

Chester started. So great had been his preoccupation that he had not given a thought to the course record. He suddenly realized now that the pro., who had done the lowest medal-score to date—the other course record was held by Peter Willard with a hundred and sixty-one, achieved in his first season—had gone out in only one better than his own figures that day.

'Sixty-eight,' he said.

'What a pity you lost those strokes at the beginning!'

'Yes,' said Chester.

He spoke absently—and, as it seemed to her, primly and without enthusiasm—for the flaming idea of having a go at the course record had only just occurred to him. Once before he had done the first nine in thirty-four, but on that occasion he had not felt that curious feeling of irresistible force which comes to a golfer at the very top of his form. Then he had been aware all the time that he had been putting chancily. They had gone in, yes, but he had uttered a prayer per putt. Today he was superior to any weak doubtings. When he tapped the ball on the green, he knew it was going to sink. The course record? Why not? What a last offering to lay at her feet? She would go away, out of his life for ever; she would marry some other bird; but the memory of that supreme round would remain with her as long as she breathed. When he won the Open and Amateur for the second—the third—the fourth time, she would say to herself, 'I was with him when he dented the record for his home course!' And he had only to pick up a couple of strokes on the last nine, to do threes at holes where he was wont to be satisfied with fours. Yes, by Vardon, he would take a whirl at it.

You, who are acquainted with these links, will no doubt say that the task which Chester Meredith had sketched out for himself—cutting two strokes off thirty-five for the second nine—was one at which Humanity might well shudder. The pro. himself, who had finished sixth in the last Open Championship, had never done better than a thirty-five, playing perfect golf and being one under par. But such was Chester's mood that, as he teed up on the tenth, he did not even consider the possibility of failure. Every muscle in his body was working in perfect co-ordination with its fellows, his wrists felt as if they were made of tempered steel, and his eyes had just that hawk-like quality which enables a man to judge his short approaches to the inch. He swung forcefully, and the ball sailed so close to the direction-post that for a moment it seemed as if it had hit it.

'Oo!' cried Felicia.

Chester did not speak. He was following the flight of the ball. It sailed over the brow of the hill, and with his knowledge of the course he could tell almost the exact patch of turf on which it must have come to rest. An iron would do the business from there, and a single putt would give him the first of the 'birdies' he required. Two minutes later he had holed out a six-foot putt for a three.

'Oo!' said Felicia again.

Chester walked to the eleventh tee in silence.

'No, never mind,' she said, as he stooped to put her ball on the sand. 'I don't think I'll play any more. I'd much rather just watch you.'

'Oh, that you could watch me through life!' said Chester, but he said it to himself. His actual words were 'Very well!' and he spoke them with a stiff coldness which chilled the girl.

The eleventh is one of the trickiest holes on the course, as no doubt you have found out for yourself. It looks absurdly simple, but that little patch of wood on the right that seems so harmless is placed just in the deadliest position to catch even the most slightly sliced drive. Chester's lacked the austere precision of his last. A hundred yards from the tee it swerved almost imperceptibly, and, striking a branch, fell in the tangled undergrowth. It took him two strokes to hack it out and put it on the green, and then his long putt, after quivering on the edge of the hole, stayed there. For a swift instant red-hot words rose to his lips, but he caught them just as they were coming out and crushed them back. He looked at his ball and looked at the hole.

'Tut!' said Chester.

Felicia uttered a deep sigh. The niblick-shot out of the rough had impressed her profoundly. If only, she felt, this superb golfer had been more human! If only she were able to be constantly in this man's society, to see exactly what it was that he did with his left wrist that gave that terrific snap to his drives, she might acquire the knack herself one of these days. For she was a clear-thinking, honest girl, and thoroughly realized that she did not get the distance she ought to with her wood. With a husband like Chester beside her to stimulate and advise, of what might she not be capable? If she got wrong in her stance, he could put her right with a word. If she had a bout of slicing, how quickly he would tell her what caused it. And she knew that she had only to speak the word to wipe out the effects of her refusal, to bring him to her side for ever.

But could a girl pay such a price? When he had got that 'eagle' on the third, he had looked bored. When he had missed this last putt, he had not seemed to care. 'Tut!' What a word to use at such a moment! No, she felt sadly, it could not be done. To marry Chester Meredith, she told herself, would be like marrying a composite of Soames Forsyte, Sir Willoughby Patterne, and all her brother Crispin's friends. She sighed and was silent.

[217]

Chester, standing on the twelfth tee, reviewed the situation swiftly, like a general before a battle. There were seven holes to play, and he had to do these in two better than bogey. The one that faced him now offered few opportunities. It was a long, slogging, dog-leg hole, and even Ray and Taylor, when they had played their exhibition game on the course, had taken fives. No opening there.

The thirteenth—up a steep hill with a long iron-shot for one's second and a blind green fringed with bunkers? Scarcely practicable to hope for better than a four. The fourteenth—into the valley with the ground sloping sharply down to the ravine? He had once done it in three, but it had been a fluke. No; on these three holes he must be content to play for a steady par and trust to picking up a stroke on the fifteenth.

The fifteenth, straightforward up to the plateau green with its circle of bunkers, presents few difficulties to the finished golfer who is on his game. A bunker meant nothing to Chester in his present conquering vein. His mashie-shot second soared almost contemptuously over the chasm and rolled to within a foot of the pin. He came to the sixteenth with the clear-cut problem before him of snipping two strokes off par on the last three holes.

To the unthinking man, not acquainted with the lay-out of our links, this would no doubt appear a tremendous feat. But the fact is, the Green Committee, with perhaps an unduly sentimental bias towards the happy ending, have arranged a comparatively easy finish to the course. The sixteenth is a perfectly plain hole with broad fairway and a down-hill run; the seventeenth, a one-shot affair with no difficulties for the man who keeps them straight; and the eighteenth, though its up-hill run makes it deceptive to the stranger and leads the unwary to take a mashie instead of a light iron for his second, has no real venom in it. Even Peter Willard has occasionally come home in a canter with a six, five, and seven, conceding himself only two eight-foot putts. It is, I think, this mild conclusion to a tough course that makes the refreshment-room of our club so noticeable for its sea of happy faces. The bar every day is crowded with rejoicing men who, forgetting the agonies of the first fifteen, are babbling of what they did on the last three. The seventeenth, with its possibilities of holing out a topped second, is particularly soothing.

Chester Meredith was not the man to top his second on any hole, so this supreme bliss did not come his way; but he laid a beautiful mashie-shot dead and got a three; and when with his iron he put his first well on the green at the seventeenth and holed out for two, life, for all his broken heart, seemed pretty tolerable. He now had the situation well in hand. He had only to play his usual game to get a four on the last and lower the course record by one stroke.

It was at this supreme moment of his life that he ran into the Wrecking Crew.

You doubtless find it difficult to understand how it came about that if the Wrecking Crew were on the course at all he had not run into them long before. The explanation is that, with a regard for the etiquette of the game unusual in these miserable men, they had for once obeyed the law that enacts that foursomes shall start at the tenth. They had begun their dark work on the second nine, accordingly, at almost the exact moment when Chester Meredith was driving off at the first, and this had enabled them to keep ahead until now. When Chester came to the eighteenth tee, they were just leaving it, moving up the fairway with their caddies in mass formation and looking to his exasperated eye like one of those great race-migrations of the Middle Ages. Wherever Chester looked he seemed to see human, so to speak, figures. One was doddering about in the long grass fifty yards from the tee, others debouched to the left and right. The course was crawling with them.

Chester sat down on the bench with a weary sigh. He knew these men. Self-centred, remorseless, deaf to all the promptings of their better nature, they never let anyone through. There was nothing to do but wait.

The Wrecking Crew scratched on. The man near the tee rolled his ball ten yards, then twenty, then thirty—he was improving. Ere long he would be out of range. Chester rose and swished his driver.

But the end was not yet. The individual operating in the rough on the left had been advancing in slow stages, and now, finding his ball teed up on a tuft of grass, he opened his shoulders and let himself go. There was a loud report, and the ball, hitting a tree squarely, bounded back almost to the tee, and all the weary work was to do again. By the time Chester was able to drive, he was reduced by impatience, and the necessity of refraining from commenting on the state of affairs as he would have wished to comment, to a frame of mind in which no man could have kept himself from pressing. He pressed, and topped. The ball skidded over the turf for a meagre hundred yards.

'D-d-d-dear me!' said Chester.

The next moment he uttered a bitter laugh. Too late a miracle had happened. One of the foul figures in front was waving its club. Other ghastly creatures were withdrawing to the side of the fairway. Now, when the harm had been done, these outcasts were signalling to him to go through. The hollow mockery of the thing swept over Chester like a wave. What was the use of going through now? He was a good three hundred yards from the green, and he needed a bogey at this hole to break the record. Almost absently he drew his brassie from his bag; then, as the full sense of his wrongs bit into his soul, he swung viciously.

Golf is a strange game. Chester had pressed on the tee and foozled. He pressed now, and achieved the most perfect shot of his life. The ball shot from its place as if a charge of powerful explosive were behind it. Never deviating from a straight line, never more than six feet from the ground, it sailed up the hill, crossed the bunker, eluded the mounds beyond, struck the turf, rolled, and stopped fifty feet from the hole. It was a brassie-shot of a lifetime, and shrill senile yippings of excitement and congratulations floated down from the Wrecking Crew. For, degraded though they were, these men were not wholly devoid of human instincts.

Chester drew a deep breath. His ordeal was over. That third shot, which would lay the ball right up to the pin, was precisely the sort of thing he did best. Almost from boyhood he had been a wizard at the short approach. He could hole out in two now on his left ear. He strode up the hill to his ball. It could not have been lying better. Two inches away there was a nasty cup in the turf; but it had avoided this and was sitting nicely perched up, smiling an invitation to the mashie-niblick. Chester shuffled his feet and eyed the flag keenly. Then he stooped to play, and Felicia watched him breathlessly. Her whole body seemed to be concentrated on him. She had forgotten everything save that she was seeing a course record get broken. She could not have been more wrapped up in his success if she had had large sums of money on it.

The Wrecking Crew, meanwhile, had come to life again. They had stopped twittering about Chester's brassie-shot and were thinking of resuming their own game. Even in foursomes where fifty yards is reckoned a good shot somebody must be away, and the man whose turn it was to play was the one who had acquired from his brother-members of the club the nickname of the First Grave-Digger.

A word about this human wen. He was—if there can be said to be grades in such a sub-species—the star performer of the Wrecking Crew. The lunches of fifty-seven years had caused his chest to slip down into the mezzanine floor, but he was still a powerful man, and had in his youth been a hammer-thrower of some repute. He differed from his colleagues—the Man With the Hoe, Old Father Time, and Consul, the Almost Human—in that, while they were content to peck cautiously at the ball, he never spared himself in his efforts to do it a violent injury. Frequently he had cut a blue dot almost in half with his niblick. He was completely muscle-bound, so that he seldom achieved anything beyond a series of chasms in the turf, but he was always trying, and it was his secret belief that, given two or three miracles happening simultaneously, he would one of these days bring off a snifter. Years of disappoint-ment had, however, reduced the flood of hope to a mere trickle, and when he took his brassie now and addressed the ball he had no immediate plans beyond a vague intention of rolling the thing a few yards farther up the hill.

The fact that he had no business to play at all till Chester had holed out did not occur to him; and even if it had occurred he would have dismissed the objection as finicking. Chester, bending over his ball, was nearly two hundred yards away—or the distance of three full brassie-shots. The First Grave-Digger did not hesitate. He whirled up his club as in distant days he had been wont to swing the hammer, and with the grunt which this performance always wrung from him, brought it down.

Golfers—and I stretch this term to include the Wrecking Crew—are a highly imitative race. The spectacle of a flubber flubbing ahead of us on the fairway inclines to make us flub as well; and, conversely, it is immediately after we have seen a magnificent shot that we are apt to eclipse ourselves. Consciously the Grave-Digger had no notion how Chester had made that superb brassie-biff of his, but all the while I suppose his subconscious self had been tak-ing notes. At any rate, on this occasion he, too, did the shot of a lifetime. As he opened his eyes, which he always shut tightly at the moment of impact, and started to unravel himself from the complicated tangle in which his follow-through had left him, he perceived the ball breasting the hill like some untamed jack-rabbit of the Californian prairie.

For a moment his only emotion was one of dreamlike amazement. He stood looking at the ball with a wholly impersonal wonder, like a man suddenly confronted with some terrific work of Nature. Then, as a sleep-walker awakens, he came to himself with a start. Directly in front of the flying ball was a man bending to make an approach-shot.

Chester, always a concentrated golfer when there was man's work to do, had scarcely heard the crack of the brassie behind him. Certainly he had paid no attention to it. His whole mind was fixed on his stroke. He measured with his eye the distance to the pin, noted the down-slope of the green, and shifted his stance a little to allow for it. Then, with a final swift waggle, he laid his club-head behind the ball and slowly raised it. It was just coming down when the world became full of shouts of 'Fore!' and something hard smote him violently on the seat of his plus fours.

The supreme tragedies of life leave us momentarily stunned. For an instant which seemed an age Chester could not understand what had happened. True, he realized that there had been an earthquake, a cloud-burst, and a railway accident, and that a high building had fallen on him at the exact moment when somebody had shot him with a gun, but these happenings would account for only a small part of his sensations. He blinked several times, and rolled his eyes wildly. And it was while rolling them that he caught sight of the gesticulating Wrecking Crew on the lower slopes and found enlightenment. Simultaneously, he observed his ball only a yard and a half from where it had been when he addressed it.

Chester Meredith gave one look at his ball, one look at the flag, one look at the Wrecking Crew, one look at the sky. His lips writhed, his forehead turned vermilion. Beads of perspiration started out on his forehead. And then, with his whole soul seething like a cistern struck by a thunderbolt, he spoke.

'! ! ! ! ! ! ! ! ! ! ! ! ! ! !' cried Chester.

Dimly he was aware of a wordless exclamation from the girl beside him, but he was too distraught to think of her now. It was as if all the oaths pent up within his bosom for so many weary days were struggling and jostling to see which could get out first. They cannoned into each other, they linked hands and formed parties, they got themselves all mixed up in weird vowel-sounds, the second syllable of some red-hot verb forming a temporary union with the first syllable of some blistering noun.

'___ ! ___ ! ! ___ ! ! ! ___ ! ! ! ! ___ ! ! ! ! !' cried Chester.

Felicia stood staring at him. In her eyes was the look of one who sees visions.

'***! ! ! ***! ! ! ***! ! ! ***! ! !' roared Chester, in part.

A great wave of emotion flooded over the girl. How she had misjudged this silver-tongued man! She shivered as she thought that, had this not happened, in another five minutes they would have parted for ever, sundered by seas of misunderstanding, she cold and scornful, he with all his music still within him.

'Oh, Mr Meredith!' she cried, faintly.

With a sickening abruptness Chester came to himself. It was as if somebody had poured a pint of ice-cold water down his back. He blushed vividly. He realized with horror and shame how grossly he had offended against all the canons of decency and good taste. He felt like the man in one of those 'What Is Wrong With This Picture?' things in the advertisements of the etiquette books.

'I beg—I beg your pardon!' he mumbled, humbly. 'Please please, forgive me. I should not have spoken like that.'

'You should! You should!' cried the girl, passionately. 'You should have said all that and a lot more. That awful man ruining your record round like that! Oh, why am I a poor weak woman with practically no vocabulary that's any use for anything!'

Quite suddenly, without knowing that she had moved, she found herself at his side, holding his hand.

'Oh, to think how I misjudged you!' she wailed. 'I thought you cold, stiff, formal, precise. I hated the way you sniggered when you foozled a shot. I see it all now! You were keeping it in for my sake. Can you ever forgive me?'

Chester, as I have said, was not a very quick-minded young man, but it would have taken a duller youth than he to fail to read the message in the girl's eyes, to miss the meaning of the pressure of her hand on his.

'My gosh!' he exclaimed wildly. 'Do you mean——? Do you think——? Do you really——? Honestly, has this made a difference? Is there any chance for a fellow, I mean?'

Her eyes helped him on. He felt suddenly confident and masterful.

[223]

'Look here—no kidding—will you marry me?' he said.

'I will! I will!'

'Darling!' cried Chester.

He would have said more, but at this point he was interrupted by the arrival of the Wrecking Crew who panted up full of apologies; and Chester, as he eyed them, thought that he had never seen a nicer, cheerier, pleasanter lot of fellows in his life. His heart warmed to them. He made a mental resolve to hunt them up some time and have a good long talk. He waved the Grave-Digger's remorse airily aside.

'Don't mention it,' he said. 'Not at all. Faults on both sides. By the way, my *fiancée*, Miss Blakeney.'

The Wrecking Crew puffed acknowledgment.

'But, my dear fellow,' said the Grave-Digger, 'it was—really it was—unforgivable. Spoiling your shot. Never dreamed I would send the ball that distance. Lucky you weren't playing an important match.'

'But he was,' moaned Felicia. 'He was trying for the course-record, and now he can't break it.'

The Wrecking Crew paled behind their whiskers, aghast at this tragedy, but Chester, glowing with the yeasty intoxication of love, laughed lightly.

'What do you mean, can't break it?' he cried, cheerily. 'I've one more shot.'

And, carelessly addressing the ball, he holed out with a light flick of his mashie-niblick.

'Chester, darling!' said Felicia.

They were walking slowly through a secluded glade in the quiet evenfall.

'Yes, precious?'

Felicia hesitated. What she was going to say would hurt him, she knew, and her love was so great that to hurt him was agony.

'Do you think——' she began. 'I wonder whether——It's about Crispin.'

'Good old Crispin!'

Felicia sighed, but the matter was too vital to be shirked. Cost what it might, she must speak her mind.

'Chester, darling, when we are married, would you mind very, *very* much if we didn't have Crispin with us *all* the time?'

Chester started.

'Good Lord!' he exclaimed. 'Don't you like him?'

'Not very much,' confessed Felicia. 'I don't think I'm clever enough for him. I've rather disliked him ever since we were children. But I know what a friend he is of yours——'

Chester uttered a joyous laugh.

'Friend of mine! Why, I can't stand the blighter! I loathe the worm! I abominate the excrescence! I only pretended we were friends because I thought it would put me in solid with you. The man is a pest and should have been strangled at birth. At school I used to kick him every time I saw him. If your brother Crispin tries so much as to set foot across the threshold of our little home, I'll set the dog on him.'

'Darling!' whispered Felicia. 'We shall be very, very happy.' She drew her arm through his. 'Tell me, dearest,' she murmured, 'all about how you used to kick Crispin at school.'

And together they wandered off into the sunset.

P. G. WODEHOUSE, Chester Forgets Himself, 1973

A. P. Herbert obviously shared Wodehouse's views.

IS A GOLFER A GENTLEMAN?

Rex v. *Haddock*

Before the Stipendiary

This case, which raised an interesting point of law upon the meaning of the word 'gentleman', was concluded today.

The Stipendiary, giving judgment, said: 'In this case the defendant, Mr Albert Haddock, is charged, under the Profane Oaths Act, 1745, with swearing and cursing on a Cornish golf-course, The penalty under the Act is a fine of one shilling for every day-labourer, soldier or seaman, two shillings for every other person under the degree of gentleman, and five shillings for every person of or above the degree of gentleman—a remarkable but not, unfortunately, unique example of a statute which lays down one law for the rich and another (more lenient) for the poor. The fine, it is clear, is leviable not upon the string or succession of oaths, but upon each individual malediction (see *Reg.* v. *Scott*, (1863) 33 L.J.M. 15). The curses charged, and admitted, in this case are over four hundred in number, and we are asked by the prosecution to inflict

a fine of one hundred pounds, assessed on the highest or gentleman's rate at five shillings a swear. The defendant admits the offences
but contends that the fine is excessive and wrongly calculated, on
the curious ground that he is not a gentleman when he is playing
golf.

'He has reminded us, in an able argument, that the law takes
notice, in many cases, of such exceptional circumstances as will
break down the normal restraints of a civilised citizen and so powerfully inflame his passions that it would be unjust and idle to apply
to his conduct the ordinary standards of the law, as for example
where without warning or preparation he discovers another man
in the act of molesting his wife or family. The law recognises that
under such provocation a reasonable man ceases for the time being
to be a reasonable man; and the defendant maintains that in the
special circumstances of his offence a gentleman ceases to be a
gentleman and should not be judged or punished as such.

'Now what were these circumstances? Broadly speaking, they
were the 12th hole on the —— golf-course, with which most of us
in this court are familiar. At that hole the player drives (or does
not drive) over an inlet of the sea, which is enclosed by cliffs some
sixty feet high. The defendant has told us that he never drives over,
but always into, this inlet or Chasm, as it is locally named. A moderate if not sensational player on other sections of the course,
before this obstacle his normal powers invariably desert him. This,
he tells us, has preyed upon his mind; he has registered, it appears,
a kind of a vow, and year after year, at Easter and in August, he
returns to this county, determined ultimately to overcome the
Chasm.

'Meanwhile, unfortunately, his tenacity has become notorious. It
is the normal procedure, it appears, if a ball is struck into the
Chasm, to strike a second, and, if that should have no better fate,
to abandon the hole. The defendant tells us that in the past he has
struck no fewer than six or seven balls in this way, some rolling
gently over the cliff and some flying far and high out to sea. But
recently, grown fatalistic, he has not thought it worth while to
make even a second attempt, but has immediately followed his first
ball into the Chasm, and there, among the rocks, small stones and
shingle, has hacked at his ball with the appropriate instrument until
some lucky blow has lofted it on to the turf above, or, in the alternative, until he has broken his instruments or suffered some injury

from flying fragments of rock. On one or two occasions a crowd of holiday-makers and local residents has gathered on the cliff and foreshore to watch the defendant's indomitable struggles and to hear the verbal observations which have accompanied them. On the date of the alleged offences a crowd collected of unprecedented dimensions, but so intense was the defendant's concentration that he did not, he tells us, notice their presence. His ball had more nearly traversed the gulf than ever before; it struck the opposing cliff but a few feet from the summit, and nothing but an adverse gale of exceptional ferocity prevented success. The defendant therefore, as he conducted his customary excavations among the boulders of the Chasm, was possessed, he tells us, by a more than customary fury. Oblivious of his surroundings, conscious only of the will to win, for fifteen or twenty minutes he lashed his battered ball against the stubborn cliffs until at last it triumphantly escaped. And before, during, and after every stroke he uttered a number of imprecations of a complex character which were carefully recorded by an assiduous caddie and by one or two of the spectators. The defendant says that he recalls with shame a few of the expressions which he used, that he has never used them before, and that it was a shock to him to hear them issuing from his own lips; and he says quite frankly than no gentleman would use such language.

'Now this ingenious defence, whatever may be its legal value, has at least some support in the facts of human experience. I am a golf-player myself—(Laughter)—but, apart from that, evidence has been called to show the subversive effect of this exercise upon the ethical and moral systems of the mildest of mankind. Elderly gentlemen, gentle in all respects, kind to animals, beloved by children and fond of music, are found in lonely corners of the Downs hacking at sand-pits or tussocks of grass and muttering in a blind ungovernable fury elaborate maledictions which could not be extracted from them by robbery with violence. Men who would face torture without a word become blasphemous at the short fourteenth. And it is clear that the game of golf may well be included in that category of intolerable provocations which may legally excuse or mitigate behaviour which is not otherwise excusable, and that under that provocation the reasonable or gentle man may reasonably act like a lunatic or lout, and should be judged as such.

'But then I have to ask myself, What does the Act intend by the words *"of or above the degree of gentleman"*? Does it intend a fixed social rank or a general habit of behaviour? In other words, is a gentleman legally always a gentleman, as a duke or a solicitor remains unalterably a duke or a solicitor? For if this is the case the defendant's argument must fail. The prosecution say that the word "degree" is used in the sense of "rank". Mr Haddock argues that it is used in the sense of an university examination, and that, like the examiners, the Legislature divides the human race, for the purposes of swearing, into three vague intellectual or moral categories, of which they give certain rough but not infallible examples. Many a First-Class man has taken a Third, and many a day-labourer, according to Mr Haddock, is of such a high character that under the Act he should rightly be included in the First "degree". There is certainly abundant judicial and literary authority for the view that by "gentleman" we mean a personal quality and not a social status. We have all heard of "Nature's gentlemen". "Clothes do not make the gentleman," said Lord Arrowroot in *Cook* v. *The Mersey Docks and Harbour Board*, (1897) 2 Q. B., meaning that a true gentleman might be clad in the foul rags of an author. In the old maxim "Manners makyth man" (see *Charles* v. *The Great Western Railway*), there is no doubt that by "man" is meant "gentleman", and that "manners" is contrasted with wealth or station. Mr Thomas, for the prosecution, has quoted against these authorities an observation of the poet Shakespeare that

The Prince of Darkness is a gentleman,

but quotations from Shakespeare are generally meaningless and always unsound. This one, in my judgment, is both. I am more impressed by the saying of another author (whose name I forget) that the King can make a nobleman, but he cannot make a gentleman.

'I am satisfied therefore that the argument of the defendant has substance. Just as the reasonable man who discovers his consort in the embraces of the supplanter becomes for the moment a raving maniac, so the habitually gentle man may become in a bunker a violent unmannerly oaf. In each case the ordinary sanctions of the law are suspended; and, while it is right that a normally gentle person should in normal circumstances suffer a heavier penalty for needless imprecations than a common seaman or cattle-driver, for whom they are part of the tools of his trade, he must not be judged by the standards of the gentle in such special circumstances as provoked the defendant.

'That provocation was so exceptional that I cannot think it was contemplated by the framers of the Act; and had golf at that date been a popular exercise I have no doubt that it would have been dealt with under a special section. I find therefore that this case is not governed by the Act. I find that the defendant at the time was not in law responsible for his actions or his speech, and I am unable to punish him in any way. For his conduct in the Chasm he will be formally convicted of Attempted Suicide while Temporarily Insane, but he leaves the court without a stain upon his character.'

A. P. HERBERT, *Punch*, 1927

❖

THE WOODEN PUTTER

It was not for want of clubs that Mr Polwinkle's handicap obstinately refused to fall below sixteen. His rack full of them extended round three sides of the smoking room. In addition, there was an enormous box resembling a sarcophagus on the floor, and in one corner was a large loose heap of clubs. To get one out of the heap without sending the others crashing to the ground was as delicate and difficult as a game of spillikins, and the housemaid had bestowed on it many an early morning malediction.

The rack along one side of the wall was clearly of a peculiarly sacred character. The clips holding the clubs were of plush, and behind each clip there was pasted on the wall an inscription in Mr Polwinkle's meticulously neat handwriting. There was a driver stated to have belonged to the great James Braid; a mashie of J. H. Taylor's; a spoon of Herd's.

Nor were illustrious amateurs unrepresented. Indeed, these were the greatest treasures in Mr Polwinkle's collection, because they had been harder to come by. The midiron had quite a long pedigree, passing through a number of obscure and intermediate stages, and ending in a blaze of glory with the awful name of Mr John Ball, who was alleged once to have played a shot with it at the request of an admirer. A putting cleek with a rather long, old-fashioned head and a battered grip bore the scrupulous inscription: ATTRIBUTED TO THE LATE MR F. G. TAIT.

Mr Polwinkle always sighed when he came to that cleek. Its authenticity was, he had to admit, doubtful. There were so many

Freddie Tait putters. Half the clubhouses in England seemed to possess one; they could hardly all be genuine. His Hilton he no longer even pretended to believe in.

'I bought that,' he would say, 'when I was a very young collector, and I'm afraid I was imposed upon.' But, at any rate, there was no doubt about his latest acquisition, before which he now paused lovingly. Here was the whole story, written down by a man, who knew another man, who knew the people with whom Mr Wethered had been staying. Mr Wethered had overslept himself, packed up his clubs in a hurry, and left his iron behind; so he had borrowed this one, and had graciously remarked that it was a very nice one.

It must not be supposed that Mr Polwinkle was ever so daring as to play with these sacred clubs. He contented himself with gazing and, on rare occasions, with a reverent waggle.

Mr Polwinkle, as I have said, was not a good player. He was aware of not playing consistently up to his sixteen handicap. If he did not always insist on his rights of giving two strokes to his friend Buffery, he might, he was conscious, have suffered the indignity of being beaten level by an eighteen handicap player; and with all this nonsense about scratch scores and a raising of the standard, he saw before him the horrid certainty of soon being eighteen himself.

This evening he was feeling particularly depressed. It had been a bad day. Buffery had won by five and four without using either of his strokes, and had hinted pretty strongly that he did not propose to accept them any more. Confound the tactless creature!

Mr Polwinkle tried to soothe himself by looking at his treasures. Ah! If only he could just for one day be endued with the slash and power of those who had played with them. If only something of their virtue could have passed into their clubs, what a splendid heritage! Such a miracle might even be possible if he had but faith enough. Coué-suggestion—better and better and better—how wonderful it would be!

Suddenly he felt a glow of new hope and inspiration. Greatly daring, he took from the rack the driver WITH WHICH as the inscription lyrically proclaimed JAMES BRAID WON THE CHAMPIONSHIP AT PRESTWICK IN 1908, WITH THE UNEXAMPLED SCORE OF 291; EIGHT STROKES BETTER THAN THE SECOND SCORE, AND PLAYING SUCH GOLF AS HAD NEVER BEEN SEEN BEFORE ON THAT CLASSIC COURSE.

He took one glance to see that his feet were in the right place—long practice enabled him to judge to an inch the position in which the furniture was safe—and then he swung.

Gracious goodness! What had happened? Back went the club, instinct with speed and power, and he felt a violent and unaccustomed wrenching round of his hips. Down it came more swiftly than ever, his knees seemed to crumple under him with the vehemence of the blow, and swish went the clubhead, right out and round in a glorious finish. A shower of glass fell all over him and he was left in darkness.

Never had he experienced anything before in the least like that tremendous sensation; the electric light had always been perfectly safe. With trembling fingers he struck a match and groped his way, crunching glass as he walked, to the two candles on the chimney piece. Once more he swung the club up; then paused at the top of the swing, as he had done so many hundreds of times before, and gazed at himself in the glass. Could it really be?

He rushed to the bookshelf, tore down *Advanced Golf*, turned to the appropriate page, and again allowed the club to swing and wrench him in its grip. There could be no doubt about it. Allowing for differences of form and feature he was Braid to the very life—the poise, the turn of the body, the very knuckles—all were the same.

The miracle had happened with one club. Would it happen with all? Out came the Taylor mashie from the rack. As he picked it up his head seemed to shake formidably, his wrists felt suddenly as if they were made of whipcord, his boots seemed to swell and clutch the ground; another second—crash!—down came the club and out came a divot of carpet, hurtling across the room, while Mr Polwinkle's eyes were fixed in a burning and furious gaze on the gaping rent that was left.

Then it really was all right. If he could swing the club like the great masters, he could surely hit the ball like them, and the next time he played Buffery, by Jove, it would not be only two strokes he could give him.

He was in the middle of being Mr Wethered when the door opened and Buffery walked in. Mr Polwinkle had got his feet so wide apart in his admirable impersonation that he could not move; for a perceptible moment he could only straddle and stare.

'They told me you were in, old chap,' began Buffery, 'so I just

walked in. What on earth are you at? I always said that light would
get it in the neck some day!' Buffery's heartiness, though well
meant, was sometimes hard to bear. 'However,' he went on, while
Mr Polwinkle was still speechless, 'what I came about was this.
You remember you said you'd come down to Sandwich with me
some day. Well, I suddenly find I can get off for three days. Will
you come?'

Mr Polwinkle hesitated a moment. He did not feel very kindly
disposed toward Buffery. He should like to practice his new styles
a little before crushing him; but still, Sandwich! And he had never
seen it.

'All right,' he said; 'I'll come!'

'Topping!' cried Buffery. 'We'll have some great matches, and I'm
going to beat you level—you see if I don't!'

Mr Polwinkle gathered himself together for an effort.

'I will give you,' he said slowly and distinctly, 'a stroke a hole,
and I'll play you for'—and he hesitated on the brink of something
still wilder—'five pounds!'

Buffery guffawed with laughter. He had never heard Mr Polwinkle
make so good a joke before.

The next evening saw them safely arrived and installed at the Bell.

The journey, though slow, had been for Mr Polwinkle full of
romance. When he changed at Minster he snuffed the air and
thought that already he could smell the sea. His mind was a jum-
ble of old championships and of the wondrous shots he was going
to play on the morrow. At dinner he managed to make Buffery
understand that he really did mean to give him a stroke a hole.
And Buffery, when at last convinced that it was not a joke, merely observed that a fiver would be a pleasant little help toward his
expenses.

After dinner he felt too restless and excited to sit still, and leaving Buffery to play bridge, wandered stealthily into the hall to see
if his precious clubs were safe. He felt a momentary shiver of horror when he found someone examining his bag. Had news of the
match been spread abroad? Was this a backer of Buffery's tamper-
ing with his clubs?

No; he appeared a harmless, friendly creature, and apologized
very nicely. He was merely, he said, amusing himself by looking at
the different sets of clubs.

'You've got some jolly good ones,' he went on, making Mr Polwinkle blush with pleasure. 'And look here, your mashie and mine might be twins—they're as like as two peas!' And he produced his own from a neighboring bag. They certainly were exactly alike; both bore the signature of their great maker; in weight and balance they were identical.

'Taylor used to play with mine himself!' said Mr Polwinkle in a voice of pride and awe. 'And this is Herd's spoon, and here's a putter of—'

'I expect he'd have played just as well with mine,' cut in the stranger—Jones was the unobtrusive name on his bag—with regrettable flippancy. 'Anyhow, they're both good clubs. Wish I could play like Taylor with mine. Well, I'm going to turn in early—good night!'

Mr Polwinkle, a little sad that Jones did not want to hear all about his collection, fastened up his bag, and thought he would go to bed, too. He lay awake for some time, for the cocks crow as persistently by night in the town of Sandwich as the larks sing by day upon the links; moreover, he was a little excited. Still, he slept at last, and dreamed of mashie shots with so much backspin on them that they pitched on Prince's and came back into the hole on St George's.

'Well,' said Buffery, as they stood next morning on the first tee at St George's, 'it's your honor—you're the giver of strokes,' he added in a rather bitter tone.

Mr Polwinkle took out the Braid driver with as nonchalant an air as he could muster. He could not help feeling horribly frightened, but no doubt the club would help him through. He gave one waggle with that menacing little shake of the club that Walton Heath knows so well, and then the ball sped away an incredible distance. It was far over the 'kitchen,' that grassy hollow that has caught and stopped so many hundreds of balls; but it had a decided hook on it, and ran on and on till it finished in the rough on the left.

One of the caddies gave a prolonged whistle of surprise and admiration. Who was this new, unknown, and infinitely mild-looking champion who made the club hum through the air like a hornet? Buffery, too, was palpably taken aback.

'I say, old chap,' he remarked, 'you seem to have been putting a lot on to your drive. Was that what you had up your sleeve?'

However, he managed to hit a very decent shot himself into the kitchen, and then, narrowly escaping that trappy little bunker on the right with his second, lay in a good strategic position in front of the big cross bunker.

Meanwhile, Mr Polwinkle was following up his own vast tee shot in an agitated state of mind. Of course, he reflected, Braid *can* hook. It was, he had read, the one human weakness to which the great man was occasionally prone, but it seemed hard that this should be the occasion. The ball lay very heavy in the rough, and worse than all he had only his own niblick, with which he was singularly ineffective. He had once had the chance of acquiring a genuine Ray, but niblicks were clumsy, ugly things and did not interest him. Why had he been such a fool?

His first effort was a lamentable top, his second only just got the ball out of the rough, with a gaping wound in its vitals. Still, there was a hope if Herd's spoon would behave itself as it should, and he addressed himself to the shot with a desperate composure.

Heavens, what was the matter with him? Was he never going to hit the ball? He felt himself growing dizzy with all those waggles, a fierce little glance at the hole between each of them. There could be no possible doubt that this spoon was a genuine Herd. Just as he felt that he must scream if it went on much longer, up went the club, and away went the ball—the most divine spoon shot ever seen—cut up into the wind to perfection; the ball pitched over the bunker, gave a dying kick or two, and lay within a yard of the hole.

Even the ranks of Tuscany could scarce forbear to cheer. 'Good shot!' growled Buffery grudgingly.

That was four—he would be down in five. The enemy with his stroke had three for the hole, but the big cross bunker yawned between him and the green. Drat the man, he had not topped it. He had pitched well over, and his approach putt lay so dead that Mr Polwinkle, though in no generous mood, had to give it to him. One down.

At the second hole at Sandwich, as all the world knows, there is a long and joyous carry from the tee. A really fine shot will soar over the bunker and the hilltop beyond, and the ball will lie in a little green valley, to be pitched home on to the green; but the short driver must make a wide tack to the right and will have a more difficult second.

Buffery, inspired by his previous win, despite his opponent's mighty drive, decided to 'go for it.' And plump went his ball into the bunker.

The Braid driver was on its best behavior this time—a magnificent shot, straight as an arrow and far over the hill.

'H'm!' said Buffery, looking discontentedly at the face of his driver. 'Is that any new patent kind of ball you are playing with?'

'No,' returned Mr Polwinkle frigidly. 'You can weigh it after the round if you like.' And they walked on in stony silence.

Buffery had to hack his ball out backward, and his third was away to the right of the green.

'Just a little flick with the mashie, sir,' said Mr Polwinkle's caddie, putting the club in his hand.

He took the mashie, but somehow he did not feel comfortable. He shifted and wriggled, and finally his eye was high in the heavens long before the ball was struck. When he looked down to earth again he found the ball had only moved about three yards forward—a total and ignominious fluff. He tried again; another fluff moved it forward but a few painful inches; again, and a third precisely similar shot deposited it in the bunker in front of his nose. Then he went berserk with his niblick, irretrievably ruined a second new ball, and gave up the hole.

'Let me look at that mashie!' he said to his caddie as he walked on toward the next tee. And, after microscopically examining its head, 'I see what it is!' he exclaimed, in frantic accents. 'It's that fellow—what's his damned name, who was looking at my clubs last night—he's mixed them up—he's got my mashie and I've got his! Do you know Mr Jones by sight?' And he turned to his caddie.

'Yes, sir. I knows him. And that's a funny thing if you've got his mashie. I was just thinking to myself that them shots of yours was just like what he plays. "Joneses," his friends call them. He'll play like a blooming pro, for a bit, and then fluff two or three—'

'Where is he now? Is he in front of us?' Mr Polwinkle interrupted. Yes, Jones had started some time ago.

'Then run as hard as you can and tell him I'm playing an important match and insist on having my mashie back. Quick now, run!'— as the caddie was going to say something. 'I'll carry the clubs!' And the caddie disappeared reluctantly in the sandhills.

'Bad luck, old man!' said Buffery, his complacency restored by that wonderfully soothing medicine of two holes up, 'But I'll tell you where to go. Now this is the Sahara. The hole's over there,' pointing to the left, 'but it's too long a carry for you and me—we must go round by the right.'

'Which line would Braid take?' asked Mr Polwinkle. 'Straight at the flag, would he? Then I shall go straight for the flag!'

'Please yourself!' answered Buffery with a shrug, and played away to the right—a mild little shot and rather sliced, but still clear of the sand. Mr Polwinkle followed with another superb tee shot. Far over all that tumultuous mass of rolling sandhills the ball flew, and was last seen swooping down on to the green. Buffery's second was weak and caught in the hollow; his third was half topped and ran well past; his fourth put him within a yard or so of the hole.

The best he could do would be a five, and all the while there stood Mr Polwinkle, calm, silent, and majestic, six yards from the flag in one. He had only to get down in two putts to win the hole; but he had not yet had a putt, and which putter was he to use— the Tait or the Harry Vardon? He decided on the Tait. A moment later he wished he had not, for his putt was the feeblest imaginable, and the ball finished a good five feet short. Still he persevered, and again was pitifully short.

'By Jove, that's a let-off, old chap!' said the tactless one, and popped his own ball into the hole.

'I'll give you that one!' he added magnanimously, and picked up Mr Polwinkle's ball, which was reposing some three inches from the hole.

'I was always afraid it was a forgery!' murmured Mr Polwinkle, mechanically accepting the ball. 'Freddie Tait was never short with his putts—the books all say that!'

Buffery looked at him wonderingly, opened his mouth as if to make some jocular comment, then thought better of it and led the way to the tee.

Much the same thing happened at the fourth. Two magnificent shots by Braid and Herd respectively, right up to the edge of the little plateau, where it stands defiantly with the black railings in the background; a series of four scrambles and scuffles by Buffery, which just escaped perdition. Two for the hole again, and this time the Vardon putter was tried. The first putt was beautiful. How sweetly and smoothly and with what a free wrist it was taken back!

The ball, perfectly struck, seemed in, then it just slipped past and lay two feet away.

'Ah!' he said to himself with a long sigh of satisfaction, 'at any rate this is genuine!'

Alas! It was but too true, for when it came to the short putt, Mr Polwinkle's wrist seemed suddenly to become locked, there was a quick little jerk of the club and—yes, somehow or other the ball had missed the hole. Buffery was down in his two putts again, and it was another half, this time in five to six.

'I ought to have been all square by now if I could have putted as well as an old lady with a broomstick!' said poor Mr Polwinkle.

'Well, I like that!' answered the other truculently. 'I ought to have been four up if I could have played a decent second either time!' And this time there was a lasting silence.

Mr Polwinkle felt depressed and miserable. Still his heart rose a little when he contemplated the bunker that had to be carried from the tee at the fifth, and beyond it the formidable Maiden with its black terraces. And, sure enough, Buffery got into the bunker in three—not into the black terraces, because, sad to say, men do not now play over the Maiden's crown, but only over the lower spurs—touching, as it were, but the skirts of her sandy garment. Still, he was in the bunker, and Mr Polwinkle had only a pitch to reach the green. Here it was that he wanted a good caddie to put an iron in his hand—to put anything there but the mashie that had played him false. But Mr Polwinkle was flustered.

'After all,' he thought, 'a mashie is a mashie, even if it is not a genuine Taylor, and if I keep my eye on the ball—'

Clean off the socket this time the ball flew away toward cover point, and buried itself in a clump of bents. Why did he not 'deem it unplayable'? I do not know. But since Mr Horace Hutchinson once ruined a medal round and probably lost the St George's Vase at the Maiden by forgetting that he could tee and lose two, Mr Polwinkle may be forgiven. When his ball ultimately emerged from the bents he had played five; they holed out in nine apiece, for Buffery had also had his adventures and the stroke settled in. Three down.

Worse was to come, for at the sixth Buffery had the impudence to get a three—a perfect tee shot and two putts; no one could give a stroke to that. At the seventh Mr Polwinkle, club in hand, walked

forward with elaborate care to survey the ground, walked backward, his eye still fixed on the green—and heeled his ball smartly
backward like a rugby forward. For a moment he was bewildered.
Then he looked at his club. His Wethered iron! Of course. It was
the tragedy of the Open Championship at St Andrews over again!

At Hades his Vardon putter again misbehaved at short range, and
Mr Polwinkle looked at it reproachfully.

'I always thought it belonged to a bad period!' he groaned, remem-
bering some of those tragic years in which the greatest of all golfers
could do everything but hole a yard putt. He would use the Vardon
no more. But, then, what on earth was he to putt with? He tried
the pseudo-Tait again at the ninth, and by dint of taking only three
putts got a half; but still he was six down.

There was one ray of comfort. There was his caddie waiting for
him, having no doubt run the villain Jones to earth, and under this
arm protruded the handle of a club.

'Well,' he shouted, 'have you got it?'

'No, sir,' the caddie answered—and embarrassment and amusement seemed to struggle together in his voice. 'Mr Jones says he's
playing an important match, too, and as you didn't send back his
mashie he's going on with yours. Said they were just the same, he
did, and he wouldn't know any difference between yours and his
own.'

'Then what's that club you've got there?' demanded Mr Polwinkle.

'The gentleman lent you this to make up, so he said,' the
caddie replied, producing a wooden putter. 'I was particularly to
tell you it belonged to someone who used it in a great match, and
blessed if I haven't forgotten who it was.'

Mr Polwinkle took the putter in his hand and could not disguise
from himself that it had no apparent merits of any description. The
shaft was warped, not bent in an upward curve as a well-bred wood-
en putter should be, and decidedly springy; no name whatever was
discernible on the head. Still, he badly needed a putter, and if it
had been used by an eminent hand—

'Think, man, think!' he exclaimed vehemently. 'You must remember!' But the caddie racked his brain in vain. And then—

'Really,' said Buffery, 'we can't wait all day while your caddie tries
to remember ancient history. This is the match we're thinking
about, and I'm six up!' And he drove off—a bad hook into the thick
and benty rough on the left.

And now, thank goodness, I have reached the end of Mr Polwinkle's misfortunes. The tide is about to turn. At the second shot Mr Wethered's iron, I regret to have to say, made another error. It just pulled the ball into that horrid trappy bunker that waits voraciously at the left-hand corner of the plateau green—and that after Buffery had played three and was not on the green.

Mr Polwinkle's temper had been badly shaken once or twice, and now it gave out entirely.

'Give me any dashed club you like!' he snarled, seized the first that came handy, and plunged into the bunker.

'Good sort of club to get out of a bunker with!' he said to himself, finding that he had a midiron in his hand, and then—out came the ball, as if it was the easiest thing in the world, and sat down within four yards of the hole.

How had it happened? Why, it was Mr Ball's iron—and did not the hero of Hoylake habitually pitch out of bunkers with a straight-faced iron? Of course he did—and played his ordinary pitches with it as well. What a thing it was to know history! Here at once was a magic niblick and a substitute for the mashie rolled into one. And just then his caddie smacked himself loudly and suddenly on the thigh.

'I've remembered it, sir. It was Tommy something—young Tommy, I think.'

'Young Tommy Morris?' gasped Mr Polwinkle breathlessly.

'Ah!' said the caddie. 'Morris—that was it!'

'Give me the wooden putter!' said Mr Polwinkle—and the ball rattled against the back of the tin. That was a four against Buffery's six. Down to five with eight to play.

It is a well-known fact that when golf is faultless there is next to nothing to write about it. The golfing reporter may say that So-and-So pushed his drive and pulled his second; but the real fact is that the great So-and-So was on the course with his tee shot, on the green with his second, and down in two putts—and kept on doing it. That is all the reporter need have said, but he says more because he has his living to earn. So have I; but, nevertheless, I shall not describe Mr Polwinkle's home-coming at full length. More brilliantly faultless golf never was seen. Braid drove magnificently, Mr Ball did all the pitching to perfection and even Mr Wethered behaved impeccably. As for the wooden putter, most of the putts went in, and even those that did not gave Buffery a cold shiver down his spine. What could poor eighteen-handicap Buffery do against it? He must need wilt under such an onslaught. If he did a respectable five, Mr Polwinkle did a 'birdie' three. If he did a long hole in six, as he did at the Suez Canal, that wooden putter holed one for a four.

Here, for those who know the course, are the figures of Mr Polwinkle's first eight holes coming home: four, three, three, four, four, four, two, four. That was enough. Buffery was a crushed man; hole after hole slipped away, and when he had reached the seventeenth green in eight, there was nothing for it but to give up the match. Six up at the turn and beaten by two and one!

As Mr Polwinkle walked triumphantly into the clubhouse he met Jones, and almost fell on his neck.

'My dear fellow,' he cried, 'I can't thank you enough for that putter. I holed everything. Never saw anything like it! I suppose,' he went on with a sudden desperate boldness, 'there's no chance of your selling it me, is there?'

'Oh no, I won't sell it!' began Jones.

'I knew it was too much to ask!' said Mr Polwinkle dejectedly.

'But I'll give it you with pleasure!'

'Oh, but I couldn't let you do that! Give me it for nothing—a put-ter that belonged to young Tommy—the greatest putter that ever—'

'Well, you see,' said Jones, 'I only told the caddie to tell you that because I thought it might put you on your putting. And, by George, it seems to have done it, too. Wonderful what a little confidence will do. You're perfectly welcome to the putter—I bought it in a toy shop for eighteen pence!'

Mr Polwinkle fell swooning to the floor.

BERNARD DARWIN, 1928

❖

SPICER

Then there is Spicer, or The Man Who Knows Exactly What I'm Doing Wrong.

Fortunately my game with Spicer was a Mixed Foursome, and Spicer was playing with his wife. Nothing else, I think, saved Spicer from a dreadful end. For when he is playing with his wife he has little time to devote to other people; and for some reason Mrs

Spicer refrains entirely from striking him with niblicks or pushing him over the cliff at the twelfth, as I should certainly have done if our match had been a single.

And yet how happily we started! A warm and mellow evening, and my drive went skimming over the first bunker as straight and swift as an old swallow migrating out of England on a cold wet day in August.

'Good shot,' said Spicer grudgingly, 'I didn't think you'd hit that. You dropped the right shoulder.'

'Oh!' said I carelessly, for my heart was full.

My next shot was a dream. But 'Slow back—slow back, man,' said Spicer sadly, while the ball was yet in the air. 'You didn't deserve *that*,' he added, as it came to rest within three feet of the hole.

'Thank you,' I said with dignity, for after two such shots as those I didn't propose to take advice from any old golf-bore, though he might have three wooden clubs, a Sammy and a jigger, and a patent ball-sponge in his pocket.

But none the less, when it was my turn to drive again, the canker had got me. 'Slow back,' I said to myself, as I waggled at the ball. 'No doubt the old fool was right. Slow back—and for the Lord's sake keep that right shoulder in the sky. We'll show him!'

And of course I hit the ball ten yards.

'I expected that,' said Spicer smoothly 'You were standing right in front of it.'

After that my game went to pieces. I could do nothing right. And Spicer, having destroyed me for the day, turned his attention to his wife. Mrs Spicer plays very badly, with a steady, methodical, consistent badness that commands one's admiration. She has played for ten years and she knows, and Spicer knows, and everybody knows, that she will never play any better. Yet she plays. She plays with Spicer. She is heroic.

The newspapers tell us that in America women seek divorces because their husbands go off and play golf without them. The crying need of English womanhood is some redress against the husbands who force their wives to play golf *with* them.

Mrs Spicer is a born fool, no doubt. Her ball lay about fifteen yards from a stone wall. After fingering doubtfully every club in her collection, she threw a timid glance at Spicer, who stood silent as the Sphinx, and took out her niblick. Spicer waited till she had done three preliminary waggles, and then:

'Take your mashie!' he snapped.

Mrs Spicer jumped like a shot doe and took out her mashie. After a long preparation she hit the ground very hard and the ball very gently.

'Lifting your head again,' groaned Spicer, savagely digging the ball out of a rut. 'How do you *expect* to get over if you lift your head?'

'But I *didn't* expect to get over, Cuthbert,' bleated the poor lady. 'You know I *never* expect to get over anything with my mashie.'

'Then why didn't you use your niblick?'

'But, Cuthbert, you *told* me——'

'It's no good arguing. You dropped the right shoulder, and that's all there is about it.'

'But, Cuthbert, I thought you said I *mustn't* lift the right——'

'I said you mustn't lift your *head*,' roared Spicer. 'Now try this. Take your mashie. No, take your niblick—no, not that one—your mashie-niblick—here, *this* one,' said Spicer, scattering her clubs like the cut corn upon the ground. 'It's a perfectly simple shot. Just hit your ball two yards to the right of that rabbit-hole—not too hard and not too soft. Keep your eye on the ball and let the club come right through. Don't cramp that left elbow. Slow back, right shoulder up, keep that little finger tight, and you'll be all right. There's nothing in it.'

Mrs Spicer approached the ball, trembling like a leaf, and miraculously hit it a full twenty yards.

'Um,' said Spicer, not unkindly, 'but you must keep that right heel down.'

'*Dear* Cuthbert!' whispered his wife a little later, with tears in her eyes. 'He *is* so patient with me. I know I'm terribly stupid at it, but it *is* difficult to think of so many parts of one's body at the same time, *isn't* it? It makes me feel quite *naked*.'

When she next had to drive, for a moment or two I feared for Mrs Spicer's reason. She waggled at her ball for a long, long time, so long that the whole party had the fidgets, and when it seemed that she was really about to aim a blow at it at last she did no such thing, but rested her club on the ground and stood like one in a trance—only we saw that her lips were moving.

'Right shoulder up—head down,' I caught faintly. She was repeating, like some magic incantation, the very last edition of Spicer's instructions.

Finally her brow puckered and, gazing downwards, she made curious motions with her feet; then, coming out of the trance, she murmured softly, 'What was it you said about the right heel, Cuthbert? Was it *down* or *up* it had to be?'

'Oh, hit it anyhow!' said Spicer savagely.

Thus encouraged, his lady walked right away from her ball, and, walking back again, just hit it, anyhow. The ball flew fair and far, a long way down the centre of the course, a superb shot.

'Oh, Cuthbert, *isn't* that a lovely one?' she cried, flushed with joy. '*Look* what a way it's gone!'

'Yes, it went well enough,' growled Spicer; 'but, good Heavens, you don't call that *golf!*'

Poor Mrs Spicer! She won't try that again.

A. P. HERBERT, *The Man about Town*, 1923

Although less well known than the famous account of the cricket match in England, their England, the following extract is an interesting and amusing picture of the way in which the social style of the game was changing, in the south of England at least. Donald's clubs—all wooden except for a single iron—would have been a normal set for playing with a 'feathery', and so would have been extremely old-fashioned over twenty years after the introduction of the rubber-cored Haskell.

An enormous man in a pale-blue uniform, tricked out with thick silver cords, and studded with cartwheel silver buttons, opened the door of the car and bowed Sir Ludovic and a little less impressively Donald Cameron, into the clubhouse. Donald was painfully conscious that his grey flannel trousers bagged at the knee and that his old blue 1914 golfing coat had a shine at one elbow and a hole at the other.

The moment he entered the club-house a superb spectacle met his dazzled gaze. It was not the parquet floor, on which his nail-studded shoes squeaked loudly, or the marble columns, or the voluptuous paintings on the ceilings, or the gilt-framed mirrors on the walls, or the chandeliers of a thousand crystals, or even the palms in their gilt pots and synthetic earth, that knocked him all of a heap. It was the group of golfers who were standing in front of the huge fire-place. There were purple jumpers and green jumpers and yellow jumpers and tartan jumpers; there were the biggest, the baggiest, the brightest plus-fours that ever dulled the lustre of a peacock's tail; there were the rosiest of lips, the gayest of cheeks, the flimsiest of silk stockings, the most orange of finger-nails and probably, if the truth were known, of toe-nails too; there were waves of an unbelievable permanence and lustre; there were jewels, on the men as well as on the women, and foot-long jade and amber cigarette-holders, and foot-long cigars with glistening cummerbunds; and there was laughter and gaiety and much bending, courtier-like, from the waist, and much raising of girlish, kohl-fringed eyes, and a great chattering. Donald felt like a navvy, and when, in his agitation, he dropped his clubs with a resounding clash upon the floor and every one stopped talking and looked at him, he wished he were dead. Another pale-blue-and-silver giant picked up the clubs, held them out at arm's length and examined them in disdainful astonishment—for after years of disuse they were very rusty—and said coldly, 'Clubs go into the locker-room, sir,' and Donald squeaked his way across the parquet after him, amid a profound silence.

The locker-room was full of young gentlemen who were discarding their jumpers—which certainly competed with Mr Shelley's idea of Life Staining the White Radiance of Eternity—in favour of brown leather jerkins fastened up the front with that singular arrangement which is called a zipper. Donald edged in furtively, hazily, watched the flunkey lay the clubs down upon a bench, and then fled in panic through the nearest open door and found himself suddenly in a wire-netted enclosure which was packed with a dense throng of caddies. The caddies were just as surprised by his appearance in their midst as the elegant ladies and gentlemen in the lounge had been by the fall of the clubs, and a deathly stillness once again paralysed Donald.

He backed awkwardly out of the enclosure, bouncing off caddy after caddy like a cork coming over a rock-studded sluice, and was brought up short at last by what seemed to be a caddy rooted immovably in the ground. Two desperate backward lunges failed to dislodge the obstacle and Donald turned and found it was the wall of the professional's shop. The caddies, and worse still, an exquisitely beautiful young lady with a cupid's-bow mouth and practically no skirt on at all, who had just emerged from the shop, watched him with profound interest. Scarlet in the face, he rushed past the radiant beauty, and hid himself in the darkest corner of the shop and pretended to be utterly absorbed in a driver which he picked out at random from the rack. Rather to his surprise, and greatly to his relief, no one molested him with up-to-date, go-getting salesmanship, and in a few minutes he had pulled himself together, and was able to look round and face the world.

Suddenly he gave a start. Something queer was going on inside him. He sniffed the air once, and then again, and then the half-forgotten past came rushing to him across the wasted years. The shining rows of clubs, the boxes of balls, the scent of leather and rubber and gripwax and pitch, the club-makers filing away over the vices and polishing and varnishing and splicing and binding, the casual members waggling a club here and there, the professional listening courteously to tales of apocryphal feats, all the old famil-iar scenes of his youth came back to him. It was eleven years since he had played a game of golf, thirteen years since he had bought a club. Thirteen wasted years. Dash it, thought Donald, damn it, blast it, I can't afford a new club—I don't want a new club, but I'm going to buy a new club. He spoke diffidently to one of the assistants who was passing behind him, and inquired the price of the drivers.

'It's a new lot just finished, sir,' said the assistant, 'and I'm not sure of the price. I'll ask Mr Glennie.'

Mr Glennie was the professional himself. The great man, who was talking to a member, or rather was listening to a member's grievances against his luck, a ritual which occupies a large part of a professional's working day, happened to overhear the assistant, and he said over his shoulder in the broadest of broad Scottish accents: 'They're fufty-twa shullin', and cheap at that.'

Donald started back. Two pounds twelve for a driver! Things had changed indeed since the days when the great Archie Simpson had sold him a brassy, brand-new, bright yellow, refulgent, with a lovely whippy shaft, for five shillings and ninepence.

His movement of Aberdonian horror brought him out of the dark corner into the sunlight which was streaming through the window, and it was the professional's turn to jump.

'It's Master Donald!' he exclaimed. 'Ye mind me, Master Donald—Jim Glennie, assistant that was at Glenavie to Tommy Anderson, that went to the States?'

'Glennie!' cried Donald, a subtle warm feeling suddenly invad-ing his body, and he grasped the professional's huge red hand.

'Man!' cried the latter, 'but I'm glad to see ye. How lang is't sin'

[245]

we used to ding awa at each other roon' Glenavie? Man, it must be years and years! And fit's aye deein' wi' yer game? Are ye plus sax or seeven?'

'Glennie,' said Donald sadly, 'I haven't touched a club since those old days. This is the first time I've set foot in a professional's shop since you took me that time to see Alex Marling at Balgownie the day before the War broke out.'

'Eh man, but you're a champion lost,' and the professional shook his head mournfully.

'But, Glennie,' went on Donald, 'where did you learn that fine Buchan accent? You never used to talk like that. Is it since you came south that you've picked it up?'

The big professional looked a little shamefaced and drew Donald back into the dark corner.

'It's good for trade,' he whispered in the pure English of Inverness. 'They like a Scot to be real Scottish. They think it makes a man what they call "a character." God knows why, but there it is. So I just humour them by talking like a Guild Street carter who's having a bit of back-chat with an Aberdeen fish-wife. It makes the profits something extraordinary.'

'Hi! Glennie, you old swindler,' shouted a stoutish, red-faced man who was smoking a big cigar and wearing a spectrosocopic suit of tweeds. 'How much do you want to sting me for this putter?'

'Thirty-twa shullin' and saxpence, Sir Walter,' replied Glennie over his shoulder, 'but ye'll be wastin' yer siller, for neither that club nor any ither wull bring ye doon below eighteen.'

A delighted laugh from a group of men behind Sir Walter greeted this sally.

'You see,' whispered Glennie, 'he'll buy it and he'll tell his friends that I tried to dissuade him, and they'll all agree that I'm a rare old character, and they'll come and buy too.'

'But fifty-two shillings for a driver!' said Donald. 'Do you mean to say they'll pay that?'

'Yes, of course they will. They'll pay anything so long as it's more than any other professional at any other club charges them. That's the whole secret. Those drivers there aren't a new set at all. They're the same set as I was asking forty-eight shillings for last week-end, but I heard during the week from a friend who keeps an eye open for me, that young Jock Robbie, over at Addingdale Manor, had put his drivers and brassies up from forty-six shillings to fifty, the dirty young dog. Not that I blame him. It's a new form of commercial competition, Master Donald, a sort of inverted price-cutting. Na, na, Muster Hennessey,' he broke into his trade voice again, 'ye dinna want ony new clubs. Ye're playin' brawly with yer auld yins. Still, if ye want to try yon spoon, tak it oot and play a couple of roons wi' it, and if ye dinna like it put it back.'

He turned to Donald again.

'That's a sure card down here. They always fall for it. They take the club and tell their friends that I've given it to them on trial because I'm not absolutely certain that it will suit their game, and they never bring it back. Not once. Did you say you wanted a driver, Master Donald?'

'Not at fifty-two shillings,' said Donald with a smile.

Glennie indignantly waved away the suggestion.

'You shall have your pick of the shop at cost price,' he said, and then, looking furtively round and lowering his voice until it was almost inaudible, he breathed in Donald's ear, 'Fifteen and six.'

Donald chose a beautiful driver, treading on air all the while and feeling eighteen years of age, and then Sir Ludovic Phibbs came into the shop.

'Ah! There you are, Cameron,' he said genially; 'there are only two couples in front of us now. Are you ready? Good morning, Glennie, you old shark. There's no use trying to swing the lead over Mr Cameron. He's an Aberdonian himself.'

As Donald went out, Glennie thrust a box of balls under his arm and whispered: 'For old times' sake!'

On the first tee Sir Ludovic introduced him to the other two players who were going to make up the match. One was a Mr Wollaston, a clean-shaven, intelligent, large, prosperous-looking man of about forty, and the other was a Mr Gyles, a very dark man, with a toothbrush moustache and a most impressive silence. Both were stockbrokers.

'Now,' said Sir Ludovic heartily, 'I suggest that we play a four-ball foursome, Wollaston and I against you two, on handicap, taking our strokes from the course, five bob corners, half a crown for each birdie, a dollar an eagle, a bob best ball and a bob aggregate and a bob a putt. What about that?'

'Good!' said Mr Wollaston. Mr Gyles nodded, while Donald, who had not understood a single word except the phrase 'four-ball foursome,' and that was incorrect—mumbled a feeble affirmative.

The stakes sounded enormous, and the reference to birds of the air sounded mysterious, but he obviously could not raise any objections.

When it was his turn to drive at the first tee, he selected a spot for his tee and tapped it with the toe of his driver. Nothing happened. He looked at his elderly caddy and tapped the ground again. Again nothing happened.

'Want a peg, Cameron?' called out Sir Ludovic.

'Oh no, it's much too early,' protested Donald, under the impression that he was being offered a drink. Every one laughed ecstatically at this typically Scottish flash of wit, and the elderly caddy lurched forward with a loathsome little contrivance of blue and white celluloid which he offered to his employer. Donald shuddered. They'd be giving him a rubber tee with a tassel in a minute, or lending him a golf-bag with tripod legs. He teed his ball on a pinch of sand with a dexterous twist of his fingers and thumb amid an incredulous silence.

Donald played the round in a sort of daze. After a few holes of uncertainty, much of his old skill came back, and he reeled off fairly good figures. He had a little difficulty with his elderly caddy at the beginning of the round, for, on asking that functionary to hand him 'the iron,' he received the reply:

'Which number, sir?' and the following dialogue ensued:

'Which number what?' faltered Donald.

'Which number iron?'

'Er—just the iron.'

'But it must have a number, sir.'

'Why must it?'

'All irons have numbers.'

'But I've only one.'

'Only a number one.'

'No. Only one.'

'Only one what, sir?'

'One iron!' exclaimed Donald, feeling that this music-hall turn might go on for a long time and must be already holding up the entire course.

The elderly caddy at last appreciated the deplorable state of affairs. He looked grievously shocked and said in a reverent tone:

'Mr Fumbledone has eleven.'

'Eleven what?' inquired the startled Donald.

'Eleven irons.'

After this revelation of Mr Fumbledon's greatness, Donald took 'the iron' and topped the ball hard along the ground. The caddy sighed deeply.

Throughout the game Donald never knew what the state of the match was, for the other three, who kept complicated tables upon the back of envelopes, reckoned solely in cash. Thus, when Donald once timidly asked his partner how they stood, the taciturn Mr Gyles consulted his envelope and replied shortly, after a brief calculation: 'You're up three dollars and a tanner.'

Donald did not venture to ask again, and he knew nothing more about the match until they were ranged in front of the bar in the club-room, when Sir Ludovic and Mr Wollaston put down the empty glasses which had, a moment ago, contained double pink gins, ordered a refill of the four glasses, and then handed over to the bewildered Donald the sum of one pound sixteen and six.

Lunch was an impressive affair. It was served in a large room, panelled in white and gold with a good deal of artificial marble scattered about the walls, by a staff of bewitching young ladies in black frocks, white aprons and caps, and black silk stockings. Bland wine-stewards drifted hither and thither, answering the Christian names and accepting orders, and passing them on to subordinates. Corks popped, the scent of the famous club fish-pie mingled itself with all the perfumes of Arabia and Mr Coty, smoke arose from rose-tipped cigarettes, and the rattle of knives and forks played an orchestral accompaniment to the sound of many voices, mostly silvery, like April rain, and full of girlish gaiety.

Sir Ludovic insisted on being host, and ordered Donald's half-pint of beer and double whiskies for himself and Mr Gyles. Mr Wollaston, pleading a diet and the strict orders of Carlsbad medicos, produced a bottle of Berncastler out of a small brown handbag, and polished it off in capital style.

The meal itself consisted of soup, the famous fish-pie, a fricassee of chicken, saddle of mutton or sirloin of roast beef, sweet, savoury, and cheese, topped off with four of the biggest glasses of hunting port that Donald had ever seen. Conversation at lunch was almost entirely about the dole. The party then went back to the main club-room, where Mr Wollaston firmly but humorously pushed Sir Ludovic into a very deep chair, and insisted upon taking up the running with four coffees and four double kümmels. Then after a couple of rubbers of bridge, at which Donald managed to win a few shillings, they sallied out to play a second round. The golf was only indifferent in the afternoon. Sir Ludovic complained that, owing to the recrudescence of what he mysteriously called 'the old trouble,' he was finding it very difficult to focus the ball clearly, and Mr Wollaston kept on over-swinging so violently that he fell over once and only just saved himself on several other occasions, and Mr Gyles developed a fit of socketing that soon became a menace to the course, causing, as it did, acute nervous shocks to a retired major-general whose sunlit nose only escaped by a miracle, and a bevy of beauty that was admiring, for some reason, the play of a well-known actor-manager.

So after eight holes the afternoon round was abandoned by common consent, and they walked back to the club-house for more bridge and much-needed refreshment. Donald was handed seventeen shillings as his inexplicable winnings over the eight holes. Later on, Sir Ludovic drove, or rather Sir Ludovic's chauffeur drove, Donald back to the corner of King's Road and Royal Avenue. On the way back, Sir Ludovic talked mainly about the dole.

Seated in front of the empty grate in his bed-sitting-room, Donald counted his winnings and reflected that golf had changed a great deal since he had last played it.

A. G. MACDONELL, *England, their England,* 1933

UNCLE JAMES'S GOLF MATCH

Uncle James has come to stay with his niece Molly and her husband Peter. He has invented a harness of elastic straps, attached to rings sewn into his clothes, which is designed to control and assist his swing and which he is determined to test on the course.

We got off about ten on Saturday morning—Uncle James and I. Molly had sewn the rings into his coat after dinner the night before under his expert eye; she had then superintended the connecting up in the morning after breakfast. And that completed her share of the performance. She flatly refused to accompany us to the links, on the plea of household duties. She equally flatly refused to speak to me alone, or even to meet my eye. So I placed Uncle James's bag of nineteen clubs in the car and we started.

It was a beautiful day for golf—soft, balmy, and without a breath of wind. Moreover, Uncle James was in a splendid temper.

'I shall do a good round this afternoon, Peter,' he affirmed confidently. 'Splendid device, this of mine. Tried one or two practise swings while you were getting the car.'

'Good,' I cried. With the new day had come a certain cheerful optimism, and I let the car out a bit more. 'But if I was you, Uncle James, I'd lie low about it. Don't tell anyone, and you might make a bit of money to-morrow.'

I could see the pride of the inventor struggling with the wonderful idea I had suggested. To actually beat somebody at golf! It opened a vista of possibility almost too marvellous for imagination.

'You see,' I continued craftily, 'people might belittle your game if they knew.'

I left it at that, and hoped for the best. There were quite a number of men about when we arrived at the club-house, and as Uncle James wanted to try his device, I fixed up a game for the morning. Then I showed him a hole where he could practise approach shots, and left him. It was a fatal move on my part: I ought to have known better. To leave Uncle James alone on a links—especially on Saturday morning—is asking for trouble. I got it.

The first man I saw as I came in after my round was Colonel Thresher. He was talking to the secretary, who was trying to soothe him.

'I'll look into it, Colonel,' he said mildly. 'Leave it to me.'

'But I tell you there's a madman on the links,' roared the irate officer. 'He's dug a hole on the seventeenth fairway big enough to bury a cow in.'

My heart sank; it was the seventeenth where I had left Uncle James.

'The damned man is a menace to public safety,' fumed the Colonel. 'He hits the ball backwards and through his legs. And he's using the most appalling language. Here he is, sir—here he is.'

I choked and turned round as Uncle James entered. I could see at a glance that he was no longer in a splendid temper. Far from it.

'The lies on this course are atrocious, Peter,' he cried as soon as he saw me—'positively atrocious.'

I attempted to intervene—but it was too late.

'And they won't be improved, sir,' roared the Colonel, 'by your exhibition of trench digging. Damn it—a man falling into some of those holes you've made would break his neck.'

[251]

'Confound your impertinence, sir,' began Uncle James shaking his fist in his rage. And then he paused suddenly: in mid-air, so to speak. A spasm of pain passed over his face, and a loud twanging noise came from the region of his back. The Colonel started violently, and retreated, while the secretary took two rapid paces to the rear.

'I told you he was mad,' muttered the Colonel nervously. 'He's got a musical box in his shirt.'

It was that remark that finished it, and removed the last vestige of Uncle James's self-control. To have his latest invention alluded to as a musical box turned him temporarily into a raving lunatic. And as other members drew near in awestruck silence a torrent of words in a strange tongue poured from his lips. It turned out to be some Indian dialect, of which my relative knew a smattering. Unfortunately, so did the Colonel, and he answered in the same language. I gathered later from an onlooker, who also understood the lingo, that honours were about easy, with the betting slightly on Uncle James. He'd got in first with some of the choicer terms of endearment. And then Uncle still further lost his head. He challenged the Colonel to a game that afternoon for a tenner—a challenge which that warrior immediately accepted with a sardonic laugh.

To every one else it seemed a most happy termination of the incident: to me it was the last straw. Uncle James had no more chance of beating the Colonel than I should have of beating Abe Mitchell. Not that the Colonel was a good golfer; he wasn't. But he was one of those steady players who can be relied on to go round in two or three over sevens. Which, with Uncle as his opponent, meant a victory for the Colonel by ten and eight.

However, the challenge had been given and accepted: there was nothing for it but to hope for the best. Uncle James had disappeared to wash his hands; the Colonel had been led away breathing hard, when I suddenly thought of Molly. After all, he was *her* relative.

'Is that you, Molly?' I said over the 'phone. 'Well, the worst has occurred. Your uncle has challenged old Colonel Thresher to a game this afternoon—after the combined efforts of most of the members just prevented a free fight in the smoking-room.'

I heard her choke gently. Then—'Well, that's all right, Peter.'

'It isn't,' I fumed. 'He's got no more chance of winning than— than—Don't you understand: Thresher called his invention a musical box. It came into action as they were abusing one another, and twanged. It's an affair of honour with Uncle James. And if he loses, he'll never forgive us.'

'He mustn't lose, Peter.' I thought her voice was thoughtful.

'Then I wish to heaven you'd come up and prevent it,' I said peevishly.

'I will,' she said, and I gasped. 'What ball is he using?'

'Silver Kings. Red dots. But look here, Molly, you mustn't. . . . It's for a tenner. . . . Are you there?'

She wasn't: she'd rung off. And somewhat pensively I joined Uncle James at the bar. I never quite know with Molly: she is capable of doing most peculiar things.

'I'll teach him, Peter.' He greeted me with a scowl. 'What did he say—musical box? The infernal scoundrel.'

'What was it that made the noise, Uncle James?' I asked soothingly.

'One of the longer rubbers got caught up in my braces,' he said. 'Incidentally it nipped a bit of my back. . . . Bah! Musical box. The villain.'

'Is it acting all right?' I led him towards the dining-room.

'I shall adjust it finally after lunch,' he stated.

'You don't think,' I hazarded, 'that as you haven't actually perfected it yet, it would perhaps be better to play without it.'

'Certainly not.' He glared sombrely at the back of his rival, and once again I heard him whisper: 'Musical box.'

Then we sat down to lunch. It was a silent meal and I was glad when it was over. Uncle James—that genial if eccentric individual—had departed: an infuriated and revengeful man had taken his place. And what would be the result on his disposition when he forked up ten Bradburys to the Colonel was beyond my mental scope. He was never at his best on the golf links: but this time. . . .

He disappeared for a considerable time, after consuming two glasses of our best light port, which he stated was completely unfit for human consumption, and I wandered thoughtfully towards the first tee. There was no sign of Molly, though I thought I saw the flutter of something red in the distance, which might have been her. And then the professional strolled up.

'Hear there's a tenner on Colonel Thresher's game,' he said affably.

'There is,' I answered grimly. 'Did you see his opponent playing this morning?'

'I saw the gentleman doing exercises on the seventeenth,' he said
guardedly.

'That's my uncle, Jenkins,' I cried bitterly—'or rather my wife's
uncle. Can you as a man and a golfer give me the faintest shadow
of hope that the match won't end on the tenth green?'

'Your uncle, is he,' he returned diplomatically. 'Peculiar style, sir,
hasn't he?'

'Peculiar,' I groaned. 'He'd earn a fortune on the variety stage.
By the way, you haven't seen my wife, have you?'

'Yes, sir. I thought she was playing with you. She's just bought
a couple of old remakes.'

'What brand, Jenkins?' I asked slowly.

'Red-dot silver kings. Seemed very keen on 'em, though she generally uses Dunlops.'

I turned away lest he should see my face. I had more or less
resigned myself to being cut out of Uncle James's will and to seeing his money go to a home for lost cats; but to be turned out of
the club as well for Molly's nefarious scheme was a bit over the
odds. What devilry she contemplated I did not know—I didn't even
try to guess. But not for nothing had she invested in two remake
red dots, and disappeared into the blue.

'Here they are,' said Jenkins. 'Odd sort of walk your uncle has
got, sir.'

Now Uncle James has many peculiarities, but I had never noticed
anything strange about his pedestrianism. The shock, therefore,
was all the greater. To what portion of his anatomy he had attached
his infernal machine factory I was in ignorance: but the net result
was fierce. He looked like a cross between a king penguin and a
trussed fowl suffering from an acute attack of locomotor ataxy. A
perfect bevy of members had gathered outside the club-house, and
were watching him with awed fascination: his caddy, after one fearful convulsion of laughter, had relapsed into his customary after-
luncheon hiccoughs. It was a dreadful spectacle—but worse, far
worse, was to come.

The Colonel stalked to the tee in grim silence. His face was a
little flushed: in his eyes was the light of battle.

'Ten pounds, you said, sir—I believe.'

'I will make it twenty, if you prefer,' said Uncle James loftily.

'Certainly,' snapped the Colonel, and addressed his ball.

Usually after lunch the Colonel fails to reach the fairway of the first hole. On this occasion, however, the ball flew quite a hundred yards down the middle of the course, and the Colonel stepped magnificently off the tee and proceeded to light a cigar.

The members drew closer as Uncle James advanced, and even the caddy forbore to hiccough. The moment was tense with emotion: it still lives in my memory and ever will.

'Slow back,' had said Vardon; 'follow through,' had ordered Ray. Merciful heavens! they should have seen the result of their teaching. Uncle James achieved the most wonderful wind shot of modern times.

He lifted his driver like a professional weight-lifter, and at about the same velocity. Then, his face grim with determination, he let it down again. To say that he followed through would be to damn with faint praise. The club itself finished twenty yards in front of the Colonel's ball, and Uncle James fell over backwards.

'Very good,' said the Colonel. 'But the object of the game is to get your ball into the hole—not your club.'

'Another driver, boy,' said Uncle James magnificently when he was again in a vertical position, and at that moment I felt proud of being related to him. Once more Uncle James lifted his club; once more, under the combined influence of the 'to left wrist for follow through' rubber and his inflexible determination, the club descended. And this time he hit the ball. In cricket phraseology point would have got it in the neck. As it was, the Colonel's caddy sprang into the air with a scream of fear, and got it in the stomach, whence the ball rebounded into the tee box.

'Confound it, sir!' roared the Colonel. 'That's my boy.'

'Precisely, sir,' returned Uncle James complacently. 'It is therefore my hole.'

For a moment I feared for Colonel Thresher's reason. Even Jenkins, a most phlegmatic man, retired rapidly behind the starter's box, and laid his head on a cold stone. In fact, only Uncle James seemed unperturbed. He unwound himself, twanged faintly, and started for the second tee.

'I must adjust my "right elbow in" grip, Peter,' he remarked as I trailed weakly behind him. 'It prevents me raising my club with the freedom required for a perfect swing.'

'Do you mean to say, sir'—the Colonel had at last found his voice—'that you intend to claim that hole?'

'I presume that we are playing under the rules of golf.' Uncle

James regarded him coldly. 'And the point is legislated for. Should a player's ball strike his opponent or his opponent's caddy the player wins the hole.'

'That doesn't apply to attempted murder off the tee,' howled the Colonel.

'You are not in the least degree funny, sir,' returned Uncle James still more coldly. 'In fact, I find you rather insulting. If you like, and care to forfeit the stakes, we will call the match off.'

'I'll be damned if I do,' roared the other. 'But before you drive next time, sir, I'll take precautions. I came out to play golf, not to be killed by a brass band.'

Uncle James turned white, but he controlled himself admirably. Even when he reached the second tee, and the Colonel, seizing his caddy, went to ground in a pot bunker, over the edge of which they both peered fearfully, Uncle retained his dignity.

'Straight down the middle is the line, I suppose,' he remarked to his caddy.

'Yus,' said the caddy from a range of twenty yards.

But unfortunately Uncle James did not go straight down the middle. It's a very nice five hole is our second: a drive, a full brassie and a mashie on to the green over a little hill. But you must get your drive—otherwise . . . And Uncle was otherwise.

I measured it afterwards. His driver hit the ground exactly eighteen inches behind the ball, travelling with all the force of 'to left wrist for follow through.' The shaft followed through: the head did not. It remained completely embedded in the turf.

'Have you finished?' demanded the Colonel, emerging from his dugout. Then he pointed an outraged finger at the broken head. 'This is a tee, sir, not a timber-yard. Would you be good enough to remove that foreign body before I drive?'

I removed it: I was afraid Uncle would twang again if he stooped. And then the Colonel addressed his ball. From there by easy stages, with a fine-losing hazard off a tree, it travelled out of bounds.

'Stroke and distance, I presume,' murmured Uncle. 'Boy, another driver.'

And then ensued a spectacle which almost shattered my nerves. Uncle James got stuck. He got his club up but he couldn't get it down. Both arms were wrapped round his neck, the club lay over his left shoulder pointing at the ground. And there he remained, saying the most dreadful things, and biting his sleeve.

'Posing for a statue?' asked the Colonel satirically.

'Grrr—' said Uncle, and suddenly something snapped. The club came down like a streak of lightning—there was a sweet, clear click, and even Duncan would have been satisfied with the result. Probably it was the most exquisite moment of Uncle's life. Heaven knows how it happened—certainly the performer didn't. But for the first time and—I feel tolerably confident—the last, Uncle James hit a perfect drive. It was three hundred yards if it was an inch, and the Colonel turned pale.

'That's two I've played,' said Uncle calmly. 'You play the odd, sir.'

It was then that the fighting spirit awoke in all its intensity in his opponent, and Uncle James followed him from bunker to bunker counting audibly until they came up with his drive.

'I'm playing one off ten,' he remarked genially.

'And you'll bally well play it,' snapped the Colonel.

Uncle James smiled tolerantly. 'Certainly. As you please. Boy, the wry-necked mashie.'

But it wasn't the wry-necked mashie's day in. Whatever Duncan might have thought about Uncle's drive, I don't think he'd have passed the wry-necked mashie. At the best of times it was a fearsome weapon—on this occasion it became diabolical. Turf and mud flew in all directions—only the ball remained *in statu quo*.

'That's like as we lie,' said the Colonel, as Uncle paused for breath.

'Confound you, sir—go away,' roared Uncle James, completely losing all vestige of self-control. And at that moment I saw Molly peering over the hill that guarded the green.

'The laid-back niblick, boy.' Uncle threw the wry-necked mashie into a neighbouring garden—and resumed the attack.

'Fourteen—fifteen—sixteen,' boomed the Colonel. 'Why not get a spade. . . . Ah! congratulations. You've hit the ball, even if you have sliced it out of bounds. Perhaps you'd replace some of the turf—or shall I send for a "ground under repair" notice.'

'Your shot, sir,' said Uncle thickly.

'Let me see—I'm playing one off six,' remarked the Colonel. 'And you're out of bounds.'

'I may not be.' Uncle ground his teeth. 'I may have hit a tree and bounced back. G-r-r-r!'

There was a loud tearing noise, and Uncle James started as if an asp had stung him.

'Confound you, sir,' howled the Colonel, as he topped his ball, 'will you be silent when I'm playing?'

But Uncle James was beyond aid.

'My God, Peter!' he muttered. 'I've come undone.'

It was only too true: he was twanging all over like a jazz band. Portions of india-rubber were popping out of his garments like worms on a damp green, and every now and then the back of his coat was convulsed by some internal spasm.

'Can't you take it off altogether?' I asked feverishly.

'No. I can't,' he snapped. 'The beastly thing is sewn in.'

We heard the Colonel's voice from the green.

'I have played sixteen,' he began—then he stopped with a strangled snort. And as we topped the hill we saw him staring horror-struck at the hole, his lips moving soundlessly.

'That was a lucky shot of yours, uncle,' came Molly's gentle voice from a shelter where she was knitting. 'Hit that log and bounced right back into the hole.'

And the brazen woman came across the green towards us literally staring me straight in the face.

'How does the game stand, Colonel Thresher?' she asked sweetly.

'The game, madam,' he choked. 'This isn't a game—it's an—an epidemic. He's murdered my caddy and dug a grave for him, and supplied the music—and now he's bounced into the hole.' He shook his putter in the air and faced Uncle James.

'You have that for a half,' said Uncle, dispassionately regarding a twenty-yard putt. Then he looked at the Colonel and frowned. 'What are you staring at, confound you, sir?'

But the Colonel was backing away, stealthily, muttering to himself.

'I knew it—I knew it,' he said shakingly. 'It's a monkey: the damned man's a musical monkey. He's got a tail—he's got two tails. He's got tails all over him. I've got 'em again: must have. What on earth will Maria say?'

'What the devil?' began Uncle James furiously.

'It's all right—quite all right, sir,' answered the Colonel. 'I'm not very well to-day. Touch of fever. Tails—scores of tails. Completely surrounded by tails. Some long—some short: some with loops—and some without. Great heavens! there's another just popped out of his neck. Must go and see a doctor at once. Never touch the club port again, I swear it. Never——'

Still muttering, he faded into the distance, leaving Uncle James speechless on the green.

'What the devil is the matter with the fool?' he roared when he had partially recovered his speech.

'I don't think he's very well, Uncle,' said Molly chokingly.

'But isn't he going to play any more?' demanded Uncle. 'He'd never have holed that putt, and I'd have been two up.'

'I know, dear,' said Molly, slipping her arm through his and leading him gently from the green.

'But I think he's a little upset.'

'Of course, if the man's ill,' began Uncle doubtfully.

'He is, Uncle James,' I said firmly—'a touch of the sun.' I warily dodged two long streamers trailing behind him, and took his other arm.

'What about going home for tea?'

Uncle brightened.

'That reminds me,' he murmured, 'I've just perfected a small device for automatically washing dirty cups and saucers.'

'Splendid,' I remarked, staring grimly at Molly. 'You shall try it this afternoon.'   SAPPER (H. H. MCNEILL), *Out of the Blue*, 1925

BOND v. GOLDFINGER

Although he was a keen golfer who, but for his untimely death, would have been captain of Royal St George's (here very thinly disguised as St Marks), Ian Fleming was obviously as shaky as most amateurs on the rules of golf. In fact, if Goldfinger had played the wrong ball he would have been deemed to have lost the seventeenth hole, so that with Bond throwing away the last the match would have been halved.

'Good afternoon, Blacking. All set?' The voice was casual, authoritative. 'I see there's a car outside. Not somebody looking for a game, I suppose?'

'I'm not sure, sir. It's an old member come back to have a club made up. Would you like me to ask him, sir?'

'Who is it? What's his name?'

Bond smiled grimly. He pricked his ears. He wanted to catch every inflection.

'A Mr Bond, sir.'

There was a pause. 'Bond?' The voice had not changed. It was politely interested. 'Met a fellow called Bond the other day. What's his first name?'

'James, sir.'

'Oh yes.' Now the pause was longer. 'Does he know I'm here?' Bond could sense Goldfinger's antennae probing the situation.

'He's in the workshop, sir. May have seen your car drive up.' Bond thought: Alfred's never told a lie in his life. He's not going to start now.

'Might be an idea.' Now Goldfinger's voice unbent. He wanted something from Alfred Blacking, some information. 'What sort of a game does this chap play? What's his handicap?'

'Used to be quite useful when he was a boy, sir. Haven't seen his game since then.'

'Hm.'

Bond could feel the man weighing it all up. Bond smelled that the bait was going to be taken. He reached into his bag and pulled out his driver and started rubbing down the grip with a block of shellac. Might as well look busy. A board in the shop creaked. Bond honed away industriously, his back to the open door.

'I think we've met before.' The voice from the doorway was low, neutral.

Bond looked quickly over his shoulder. 'My God, you made me jump. Why——' recognition dawned—'it's Gold, Goldman . . . er— Goldfinger.' He hoped he wasn't overplaying it. He said with a hint of dislike, or mistrust, 'Where have you sprung from?'

'I told you I played down here. Remember?' Goldfinger was looking at him shrewdly. Now the eyes opened wide. The X-ray gaze pierced through to the back of Bond's skull.

'No.'

'Did not Miss Masterton give you my message?'

'No. What was it?'

'I said I would be over here and that I would like a game of golf with you.'

'Oh, well,' Bond's voice was coldly polite, 'we must do that some day.'

'I was playing with the professional. I will play with you instead.' Goldfinger was stating a fact.

There was no doubt that Goldfinger was hooked. Now Bond must play hard to get.

'Why not some other time? I've come to order a club. Anyway

I'm not in practice. There probably isn't a caddie.' Bond was being
as rude as he could. Obviously the last thing he wanted to do was
play with Goldfinger.

'I also haven't played for some time.' (Bloody liar, thought Bond.)
'Ordering a club will not take a moment.' Goldfinger turned back
into the shop. 'Blacking, have you got a caddie for Mr Bond?'

'Yes, sir.'

'Then that is arranged.'

Bond wearily thrust his driver back into his bag. 'Well, all right
then.' He thought of a final way of putting Goldfinger off. He said
roughly, 'But I warn you I like playing for money. I can't be bothered to knock a ball round just for the fun of it.' Bond felt pleased
with the character he was building up for himself.

Was there a glint of triumph, quickly concealed, in Goldfinger's
pale eyes? He said indifferently, 'That suits me. Anything you like.
Off handicap, of course. I think you said you're nine.'

'Yes.'

Goldfinger said carefully, 'Where, may I ask?'

'Huntercombe.' Bond was also nine at Sunningdale. Huntercombe
was an easier course. Nine at Huntercombe wouldn't frighten
Goldfinger.

'And I also am nine. Here. Up on the board. So it's a level game.
Right?'

Bond shrugged. 'You'll be too good for me.'

'I doubt it. However,' Goldfinger was offhand, 'tell you what I'll
do. That bit of money you removed from me in Miami. Remember?
The big figure was ten. I like a gamble. It will be good for me to
have to try. I will play you double or quits for that.'

Bond said indifferently, 'That's too much.' Then, as if he thought
better of it, thought he might win, he said—with just the right
amount of craft mixed with reluctance—'Of course you can say
that was "found money". I won't miss it if it goes again. Oh, well,
all right. Easy come, easy go. Level match. Ten thousand dollars
it is.'

Goldfinger turned away. He said, and there was a sudden sweet-
ness in the flat voice, 'That's all arranged then, Mr Blacking. Many
thanks. Put your fee down on my account. Very sorry we shall be
missing our game. Now, let me pay the caddie fees.'

Alfred Blacking came into the workroom and picked up Bond's
clubs. He looked very directly at Bond. He said, 'Remember what

I told you, sir.' One eye closed and opened again. 'I mean about that flat swing of yours. It needs watching—all the time.'

Bond smiled at him. Alfred had long ears. He might not have caught the figure, but he knew that somehow this was to be a key game. 'Thanks, Alfred. I won't forget. Four Penfolds—with hearts on them. And a dozen tees. I won't be a minute.'

Bond walked through the shop and out to his car. The bowler-hatted man was polishing the metal work of the Rolls with a cloth. Bond felt rather than saw him stop and watch Bond take out his zip bag and go into the club house. The man had a square flat yellow face. One of the Koreans?

Bond paid his green-fee to Hampton, the steward, and went into the changing-room. It was just the same—the same tacky smell of old shoes and socks and last summer's sweat. Why was it a tradition of the most famous golf clubs that their standard of hygiene should be that of a Victorian private school? Bond changed his socks and put on the battered old pair of nailed Saxones. He took off the coat of his yellowing black and white hound's-tooth suit and pulled on a faded black wind-cheater. Cigarettes? Lighter? He was ready to go.

Bond walked slowly out, preparing his mind for the game. On purpose he had needled this man into a high, tough match so that Goldfinger's respect for him should be increased and Goldfinger's view of Bond—that he was the type of ruthless, hard adventurer who might be very useful to Goldfinger—would be confirmed. Bond had thought that perhaps a hundred-pound Nassau would be the form. But ten thousand dollars! There had probably never been such a high singles game in history—except in the finals of American Championships or in the big amateur Calcutta Sweeps where it was the backers rather than the players who had the money on. Goldfinger's private accounting must have taken a nasty dent. He wouldn't have liked that. He would be aching to get some of his money back. When Bond had talked about playing high, Goldfinger had seen his chance. So be it. But one thing was certain, for a hundred reasons Bond could not afford to lose.

He turned into the shop and picked up the balls and tees from Alfred Blacking.

'Hawker's got the clubs, sir.'

Bond strolled out across the five hundred yards of shaven seaside turf that led to the first tee. Goldfinger was practising on the putting green. His caddie stood near by, rolling balls to him. Goldfinger putted in the new fashion—between his legs with a mallet putter. Bond felt encouraged. He didn't believe in the system. He knew it was no good practising himself. His old hickory Calamity Jane had its good days and its bad. There was nothing to do about it. He knew also that the St Marks practice green bore no resemblance, in speed or texture, to the greens on the course.

Bond caught up with the limping, insouciant figure of his caddie who was sauntering along chipping at an imaginary ball with Bond's blaster. 'Afternoon, Hawker.'

'Afternoon, sir.' Hawker handed Bond the blaster and threw down three used balls. His keen sardonic poacher's face split in a wry grin of welcome. 'How've you been keepin', sir? Played any golf in the last twenty years? Can you still put them on the roof of the starter's hut?' This referred to the day when Bond, trying to do just that before a match, had put two balls through the starter's window.

'Let's see.' Bond took the blaster and hefted it in his hand, gauging the distance. The tap of the balls on the practice green had ceased. Bond addressed the ball, swung quickly, lifted his head and shanked the ball almost at right angles. He tried again. This time it was a dunch. A foot of turf flew up. The ball went ten yards. Bond turned to Hawker, who was looking his most sardonic. 'It's all right, Hawker. Those were for show. Now then, one for you.' He stepped up to the third ball, took his club back slowly and whipped the club head through. The ball soared a hundred feet, paused elegantly, dropped eighty feet on to the thatched roof of the starter's hut and bounced down.

Bond handed back the club. Hawker's eyes were thoughtful, amused. He said nothing. He pulled out the driver and handed it to Bond. They walked together to the first tee, talking about Hawker's family.

Goldfinger joined them, relaxed, impassive. Bond greeted Goldfinger's caddie, an obsequious, talkative man called Foulks whom Bond had never liked. Bond glanced at Goldfinger's clubs. They were a brand new set of American Ben Hogans with smart St Marks leather covers for the woods. The bag was one of the stitched black leather holdalls favoured by American pros. The clubs were in individual cardboard tubes for easy extraction. It was a pretentious outfit, but the best.

'Toss for honour?' Goldfinger flicked a coin.

'Tails.'

It was heads. Goldfinger took out his driver and unpeeled a new ball. He said, 'Dunlop 65. Number One. Always use the same ball. What's yours?'

'Penfold. Hearts.'

Goldfinger looked keenly at Bond. 'Strict Rules of Golf?'

'Naturally.'

'Right.' Goldfinger walked on to the tee and teed up. He took one or two careful, concentrated practice swings. It was a type of swing Bond knew well—the grooved, mechanical, repeating swing of someone who had studied the game with great care, read all the books and spent five thousand pounds on the finest pro teachers. It would be a good, scoring swing which might not collapse under pressure. Bond envied it.

Goldfinger took up his stance, waggled gracefully, took his club head back in a wide slow arc and, with his eyes glued to the ball, broke his wrists correctly. He brought the club head mechanically, effortlessly, down and through the ball and into a rather artificial, copybook finish. The ball went straight and true about two hundred yards down the fairway.

It was an excellent, uninspiring shot. Bond knew that Goldfinger would be capable of repeating the same swing with different clubs again and again round the eighteen holes.

Bond took his place, gave himself a lowish tee, addressed the ball with careful enmity and, with a flat, racket-player's swing in which there was just too much wrist for safety, lashed the ball away. It was a fine, attacking drive that landed past Goldfinger's ball and rolled on fifty yards. But it had had a shade of draw and ended on the edge of the left-hand rough.

They were two good drives. As Bond handed his club to Hawker and strolled off in the wake of the more impatient Goldfinger, he smelled the sweet smell of the beginning of a knock-down-and-drag-out game of golf on a beautiful day in May with the larks singing over the greatest seaside course in the world.

The first hole of the Royal St Marks is four hundred and fifty yards long—four hundred and fifty yards of undulating fairway with one central bunker to trap a mis-hit second shot and a chain of bunkers guarding three-quarters of the green to trap a well-hit one. You can slip through the unguarded quarter, but the fairway slopes to the right there and you are more likely to end up with a nasty first-chip-of-the-day out of the rough. Goldfinger was well placed to try for this opening. Bond watched him take what was probably a spoon, make his two practice swings and address the ball.

Many unlikely people play golf, including people who are blind, who have only one arm, or even no legs, and people often wear bizarre clothes to the game. Other golfers don't think them odd, for there are no rules of appearance or dress at golf. That is one of its minor pleasures. But Goldfinger had made an attempt to look smart at golf and that is the only way of dressing that is incongruous on a links. Everything matched in a blaze of rust-coloured tweed from the buttoned 'golfer's cap' centred on the huge, flaming red hair, to the brilliantly polished almost orange shoes. The plus-four suit was too well cut and the plus-fours themselves had been pressed down the sides. The stockings were of a matching heather mixture and had green garter tabs. It was as if Goldfinger had gone to his tailor and said, 'Dress me for golf—you know, like they wear in Scotland.' Social errors made no impression on Bond, and for the matter of that he rarely noticed them. With Goldfinger it was different. Everything about the man had grated on Bond's teeth from the first moment he had seen him. The assertive blatancy of his clothes was just part of the malevolent animal magnetism that had affected Bond from the beginning.

Goldfinger executed his mechanical, faultless swing. The ball flew true but just failed to make the slope and curled off to the right to finish pin high off the green in the short rough. Easy five. A good chip could turn it into a four, but it would have to be a good one.

Bond walked over to his ball. It was lying cocked up, just off the fairway. Bond took his number four wood. Now for the 'all air route'—a soaring shot that would carry the cross-bunkers and give him two putts for a four. Bond remembered the dictum of the pros: 'It's never too early to start winning.' He took it easy, determined not to press for the long but comfortable carry.

As soon as Bond had hit the shot he knew it wouldn't do. The difference between a good golf shot and a bad one is the same as the difference between a beautiful and a plain woman—a matter of millimetres. In this case, the club face had gone through just that one millimetre too low under the ball. The arc of flight was high and soft—no legs. Why the hell hadn't he taken a spoon or a two iron off that lie? The ball hit the lip of the far bunker and fell back. Now it was the blaster, and fighting for a half.

Bond never worried too long about his bad or stupid shots. He put them behind him and thought of the next. He came up with the bunker, took his blaster and measured the distance to the pin. Twenty yards. The ball was lying well back. Should he splash it out with a wide stance and an outside-in swing, or should he blast it and take plenty of sand? For safety's sake he would blast it out. Bond went down into the bunker. Head down and follow well through. The easiest shot in golf. Try and put it dead. The wish, half way down his back swing, hurried the hands in front of the club head. The loft was killed and there was the ball rolling back off the face. Get it out, you bloody fool, and hole a long putt! Now Bond took too much sand. He was out, but barely on the green. Goldfinger bent to his chip and kept his head down until the ball was half-way to the hole. The ball stopped three inches from the pin. Without waiting to be given the putt, Goldfinger turned his back on Bond and walked off towards the second tee. Bond picked up his ball and took his driver from Hawker.

'What does he say his handicap is, sir?'

'Nine. It's a level match. Have to do better than that though. Ought to have taken my spoon for the second.'

Hawker said encouragingly, 'It's early days yet, sir.'

Bond knew it wasn't. It was always too early to start losing.

Goldfinger had already teed up. Bond walked slowly behind him, followed by Hawker. Bond stood and leant on his driver. He said, 'I thought you said we would be playing the strict rules of golf. But I'll give you that putt. That makes you one up.'

Goldfinger nodded curtly. He went through his practice routine and hit his usual excellent, safe drive.

The second hole is a three hundred and seventy yard dog-leg to the left with deep cross-bunkers daring you to take the tiger's line. But there was a light helping breeze. For Goldfinger it would now be a five iron for his second. Bond decided to try and make it easier for himself and only have a wedge for the green. He laid his ears back and bit the ball hard and straight for the bunkers. The breeze got under the slight draw and winged the ball on and over. The ball pitched and disappeared down into the gully just short of the green. A four. Chance of a three.

Goldfinger strode off without comment. Bond lengthened his stride and caught up. 'How's the agoraphobia? Doesn't all this wide open space bother it?'

'No.'

Goldfinger deviated to the right. He glanced at the instant, half-hidden flag, planning his second shot. He took his five iron and hit a good, careful shot which took a bad kick short of the green and ran down into the thick grass to the left. Bond knew that territory. Goldfinger would be lucky to get down in two.

Bond walked up to his ball, took the wedge and flicked the ball on to the green with plenty of stop. The ball pulled up and lay a yard past the hole. Goldfinger executed a creditable pitch but missed the twelve-foot putt. Bond had two for the hole from a yard. He didn't wait to be given the hole but walked up and putted. The ball stopped an inch short. Goldfinger walked off the green. Bond knocked the ball in. All square.

The third is a blind two hundred and forty yards, all carry, a difficult three. Bond chose his brassie and hit a good one. It would be on or near the green. Goldfinger's routine drive was well hit but would probably not have enough steam to carry the last of the rough and trickle down into the saucer of the green. Sure enough, Goldfinger's ball was on top of the protecting mound of rough. He had a nasty, cuppy lie, with a tuft just behind the ball. Goldfinger stood and looked at the lie. He seemed to make up his mind. He stepped past his ball to take a club from the caddie. His left foot came down just behind the ball, flattening the tuft. Goldfinger could now take his putter. He did so and trickled the ball down the bank towards the hole. It stopped three feet short.

Bond frowned. The only remedy against a cheat at golf is not to play with him again. But that was no good in this match. Bond had no intention of playing with the man again. And it was no good starting a you-did-I-didn't argument unless he caught Goldfinger doing something even more outrageous. Bond would just have to try and beat him, cheating and all.

Now Bond's twenty-foot putt was no joke. There was no question of going for the hole. He would have to concentrate on laying it dead. As usual, when one plays to go dead, the ball stopped short—a good yard short. Bond took a lot of trouble about the putt and holed it, sweating. He knocked Goldfinger's ball away. He would go on giving Goldfinger missable putts until suddenly Bond would ask him to hole one. Then that one might look just a bit more difficult.

Still all square. The fourth is four hundred and sixty yards. You drive over one of the tallest and deepest bunkers in the United Kingdom and then have a long second shot across an undulating hilly fairway to a plateau green guarded by a final steep slope which makes it easier to take three putts than two.

Bond picked up his usual fifty yards on the drive and Goldfinger hit two of his respectable shots to the gully below the green. Bond, determined to get up, took a brassie instead of a spoon and went over the green and almost up against the boundary fence. From there he was glad to get down in three for a half.

The fifth was again a long carry, followed by Bond's favourite second shot on the course—over bunkers and through a valley between high sand-dunes to a distant, taunting flag. It is a testing hole for which the first essential is a well-placed drive. Bond stood on the tee, perched high up in the sand-hills, and paused before the shot while he gazed at the glittering distant sea and at the far-away crescent of white cliffs beyond Pegwell Bay. Then he took up his stance and visualized the tennis court of turf that was his target. He took the club back as slowly as he knew how and started down for the last terrific acceleration before the club head met the ball. There was a dull clang on his right. It was too late to stop. Desperately Bond focused the ball and tried to keep his swing all in one piece. There came the ugly clonk of a mis-hit ball. Bond's head shot up. It was a lofted hook. Would it have the legs? Get on! Get on! The ball hit the top of a mountain of rough and bounced over. Would it reach the beginning of the fairway?

Bond turned towards Goldfinger and the caddies, his eyes fierce. Goldfinger was straightening up. He met Bond's eyes indifferently. 'Sorry. Dropped my driver.'

'Don't do it again,' said Bond curtly. He stood down off the tee and handed his driver to Hawker. Hawker shook his head sympathetically. Bond took out a cigarette and lit it. Goldfinger hit his drive the dead straight regulation two hundred yards.

They walked down the hill in a silence which Goldfinger unexpectedly broke. 'What is the firm you work for?'

'Universal Export.'

'And where do they hang out?'

'London. Regent's Park.'

'What do they export?'

Bond woke up from his angry ruminations. Here, pay attention! This is work, not a game. All right, he put you off your drive, but you've got your cover to think about. Don't let him needle you into making mistakes about it. Build up your story. Bond said casually, 'Oh everything from sewing-machines to tanks.'

'What's your speciality?'

Bond could feel Goldfinger's eyes on him. He said, 'I look after the small arms side. Spend most of my time selling miscellaneous ironmongery to sheiks and rajahs—anyone the Foreign Office decides doesn't want the stuff to shoot at us with.'

'Interesting work.' Goldfinger's voice was flat, bored.

'Not very. I'm thinking of quitting. Came down here for a week's holiday to think it out. Not much future in England. Rather like the idea of Canada.'

'Indeed?'

They were past the rough and Bond was relieved to find that his ball had got a forward kick off the hill on to the fairway. The fairway curved slightly to the left and Bond had even managed to pick up a few feet on Goldfinger. It was Goldfinger to play. Goldfinger took out his spoon. He wasn't going for the green but only to get over the bunkers through the valley.

Bond waited for the usual safe shot. He looked at his own lie. Yes, he could take his brassie. There came the wooden thud of a mis-hit. Goldfinger's ball, hit off the heel, sped along the ground and into the stony wastes of Hell Bunker—the widest bunker and the only unkempt one, because of the pebbles, on the course.

For once Homer had nodded—or rather, lifted his head. Perhaps his mind had been half on what Bond had told him. Good show! But Goldfinger might still get down in three more. Bond took out his brassie. He couldn't afford to play safe. He addressed the ball, seeing in his mind's eye its eighty-eight-millimetre trajectory through the valley and then the two or three bounces that would take it on to the green. He laid off a bit to the right to allow for his draw. Now!

There came a soft clinking away to his right. Bond stood away from his ball. Goldfinger had his back to Bond. He was gazing out to sea, rapt in its contemplation, while his right hand played 'unconsciously' with the money in his pocket.

[269]

Bond smiled grimly. He said, 'Could you stop shifting bullion till after my shot?'

Goldfinger didn't turn round or answer. The noise stopped.

Bond turned back to his shot, desperately trying to clear his mind again. Now the brassie was too much of a risk. It needed too good a shot. He handed it to Hawker and took his spoon and banged the ball safely through the valley. It ran on well and stopped on the apron. A five, perhaps a four.

Goldfinger got well out of the bunker and put his chip dead. Bond putted too hard and missed the one back. Still all square.

The sixth, appropriately called 'The Virgin', is a famous short hole in the world of golf. A narrow green, almost ringed with bunkers, it can need anything from an eight to a two iron according to the wind. Today, for Bond, it was a seven. He played a soaring shot, laid off to the right for the wind to bring it in. It ended twenty feet beyond the pin with a difficult putt over and down a shoulder. Should be a three. Goldfinger took his five and played it straight. The breeze took it and it rolled into the deep bunker on the left. Good news! That would be the hell of a difficult three.

They walked in silence to the green. Bond glanced into the bunker. Goldfinger's ball was in a deep heel-mark. Bond walked over to his ball and listened to the larks. This was going to put him one up. He looked for Hawker to take his putter, but Hawker was the other side of the green, watching with intent concentration Goldfinger play his shot. Goldfinger got down into the bunker with his blaster. He jumped up to get a view of the hole and then settled himself for the shot. As his club went up Bond's heart lifted. He was going to try and flick it out—a hopeless technique from that buried lie. The only hope would have been to explode it. Down came the club, smoothly, without hurry. With hardly a handful of sand the ball curved up out of the deep bunker, bounced once and lay dead!

Bond swallowed. Blast his eyes! How the hell had Goldfinger managed that? Now, out of sour grapes, Bond must try for his two. He went for it, missed the hole by an inch and rolled a good yard past. Hell and damnation! Bond walked slowly up to the putt, knocking Goldfinger's ball away. Come on, you bloody fool! But the spectre of the big swing—from an almost certain one up to a possible one down—made Bond wish the ball into the hole instead of tapping it in. The coaxed ball, lacking decision, slid past the lip. One down!

Now Bond was angry with himself. He, and he alone, had lost that hole. He had taken three putts from twenty feet. He really must pull himself together and get going.

At the seventh, five hundred yards, they both hit good drives and Goldfinger's immaculate second lay fifty yards short of the green. Bond took his brassie. Now for the equalizer! But he hit from the top, his club head came down too far ahead of the hands and the smothered ball shot into one of the right-hand bunkers. Not a good lie, but he must put it on the green. Bond took a dangerous seven and failed to get it out. Goldfinger got his five. Two down. They halved the short eighth in three. At the ninth, Bond, determined to turn only one down, again tried to do too much off a poor lie. Goldfinger got his four to Bond's five. Three down at the turn! Not too good. Bond asked Hawker for a new ball. Hawker unwrapped it slowly, waiting for Goldfinger to walk over the hillock to the next tee. Hawker said softly, 'You saw what he did at The Virgin, sir?'

'Yes, damn him. It was an amazing shot.'

Hawker was surprised. 'Oh, you didn't see what he did in the bunker, sir?'

'No, what? I was too far away.'

The other two were out of sight over the rise. Hawker silently walked down into one of the bunkers guarding the ninth green, kicked a hole with his toe and dropped the ball in the hole. He then stood just behind the half-buried ball with his feet close together. He looked up at Bond. 'Remember he jumped up to look at the line to the hole, sir?'

'Yes.'

'Just watch this, sir.' Hawker looked towards the ninth pin and jumped, just as Goldfinger had done, as if to get the line. Then he looked up at Bond again and pointed to the ball at his feet. The heavy impact of the two feet just behind the ball had levelled the hole in which it had lain and had squeezed the ball out so that it was now perfectly teed for an easy shot—for just the easy cut-up shot which had seemed utterly impossible from Goldfinger's lie at The Virgin.

Bond looked at his caddie for a moment in silence. Then he said, 'Thanks, Hawker. Give me the bat and the ball. Somebody's going to be second in this match, and I'm damned if it's going to be me.'

'Yes, sir,' said Hawker stolidly. He limped off on the short cut that would take him half-way down the tenth fairway.

Bond sauntered slowly over the rise and down to the tenth tee. He hardly looked at Goldfinger who was standing on the tee swishing his driver impatiently. Bond was clearing his mind of everything but cold, offensive resolve. For the first time since the first tee, he felt supremely confident. All he needed was a sign from heaven and his game would catch fire.

The tenth at the Royal St Marks is the most dangerous hole on the course. The second shot, to the skiddy plateau green with cavernous bunkers to right and left and a steep hill beyond, has broken many hearts. Bond remembered that Philip Scrutton, out in four under fours in the Gold Bowl, had taken a fourteen at this hole, seven of them ping-pong shots from one bunker to another, to and fro across the green. Bond knew that Goldfinger would play his second to the apron, or short of it, and be glad to get a five. Bond must go for it and get his four.

Two good drives and, sure enough, Goldfinger well up on the apron with his second. A possible four. Bond took his seven, laid off plenty for the breeze and fired the ball off into the sky. At first he thought he had laid off too much, but then the ball began to float to the left. It pitched and stopped dead in the soft sand blown on to the green from the right-hand bunker. A nasty fifteen-foot putt. Bond would now be glad to get a half. Sure enough, Goldfinger putted up to within a yard. That, thought Bond as he squared up to his putt, he will have to hole. He hit his own putt fairly smartly to get it through the powdering of sand and was horrified to see it going like lightning across the skiddy green. God, he was going to have not a yard, but a two-yard putt back! But suddenly, as if drawn by a magnet, the ball swerved straight for the hole, hit the back of the tin, bounced up and fell into the cup with an audible rattle. The sign from heaven! Bond went up to Hawker, winked at him and took his driver.

They left the caddies and walked down the slope and back to the next tee. Goldfinger said coldly, 'That putt ought to have run off the green.'

Bond said off-handedly, 'Always give the hole a chance.' He teed up his ball and hit his best drive of the day down the breeze. Wedge and one putt? Goldfinger hit his regulation shot and they walked off again. Bond said, 'By the way, what happened to that nice Miss Masterton?'

Goldfinger looked straight in front of him. 'She left my employ.'

Bond thought, good for her! He said, 'Oh, I must get in touch with her again. Where did she go to?'

'I couldn't say.' Goldfinger walked away from Bond towards his ball. Bond's drive was out of sight, over the ridge that bisected the fairway. It wouldn't be more than fifty yards from the pin. Bond thought he knew what would be in Goldfinger's mind, what is in most golfers' minds when they smell the first scent of a good lead melting away. Bond wouldn't be surprised to see that grooved swing quicken a trifle. It did. Goldfinger hooked into a bunker on the left of the green.

Now was the moment when it would be the end of the game if Bond made a mistake, let his man off the hook. He had a slightly downhill lie, otherwise an easy chip—but to the trickiest green on the course. Bond played it like a man. The ball ended six feet from the pin. Goldfinger played well out of his bunker, but missed the longish putt. Now Bond was only one down.

They halved the dog-leg twelfth in inglorious fives and the longish thirteenth also in fives, Goldfinger having to hole a good putt to do so.

Now a tiny cleft of concentration had appeared on Goldfinger's massive, unlined forehead. He took a drink of water from the tap beside the fourteenth tee. Bond waited for him. He didn't want a sharp clang from that tin cup when it was out-of-bounds over the fence to the right and the drive into the breeze favouring a slice! Bond brought his left hand over to increase his draw and slowed down his swing. The drive, well to the left, was only just adequate, but at least it had stayed in bounds. Goldfinger, apparently unmoved by the out-of-bounds hazard, hit his standard shot. They both negotiated the transverse canal without damage and it was another half in five. Still one down and now only four to play.

The four hundred and sixty yards fifteenth is perhaps the only hole where the long hitter may hope to gain one clear shot. Two smashing woods will just get you over the line of bunkers that lie right up against the green. Goldfinger had to play short of them with his second. He could hardly improve on a five and it was up to Bond to hit a really godlike second shot from a barely adequate drive.

The sun was on its way down and the shadows of the four men were beginning to lengthen. Bond had taken up his stance. It was a good lie. He kept his driver. There was dead silence as he gave his two incisive waggles. This was going to be a vital stroke. Remember to pause at the top of the swing, come down slow and whip the club head through at the last second. Bond began to take the club back. Something moved at the corner of his right eye. From nowhere the shadow of Goldfinger's huge head approached the ball on the ground, engulfed it and moved on. Bond let his swing take itself to pieces in sections. Then he stood away from his ball and looked up. Goldfinger's feet were still moving. He was looking carefully up at the sky.

'Shades please, Goldfinger.' Bond's voice was furiously controlled.

Goldfinger stopped and looked slowly at Bond. The eyebrows were raised a fraction in inquiry. He moved back and stood still, saying nothing.

Bond went back to his ball. Now then, relax! To hell with Goldfinger. Slam that ball on to the green. Just stand still and hit it. There was a moment when the world stood still, then . . . then somehow Bond did hit it—on a low trajectory that mounted gracefully to carry the distant surf of the bunkers. The ball hit the bank below the green, bounced high with the impact and rolled out of sight into the saucer round the pin.

Hawker came up and took the driver out of Bond's hand. They walked on together. Hawker said seriously, 'That's one of the finest shots I've seen in thirty years.' He lowered his voice. 'I thought he'd fixed you then, sir.'

'He dammed nearly did, Hawker. It was Alfred Blacking that hit that ball, not me.' Bond took out his cigarettes, gave one to Hawker and lit his own. He said quietly, 'All square and three to play. We've got to watch those next three holes. Know what I mean?'

'Don't you worry, sir. I'll keep my eye on him.'

They came up with the green. Goldfinger had pitched on and had a long putt for a four, but Bond's ball was only two inches away from the hole. Goldfinger picked up his ball and walked off the green. They halved the short sixteenth in good threes. Now there were the two long holes home. Fours would win them. Bond hit a fine drive down the centre. Goldfinger pushed his far out to the right into deep rough. Bond walked along trying not to be too jubilant, trying not to count his chickens. A win for him at this hole and he would only need a half at the eighteenth for the match. He prayed that Goldfinger's ball would be unplayable or, better still, lost.

Hawker had gone ahead. He had already laid down his bag and was busily—far too busily to Bond's way of thinking—searching for Goldfinger's ball when they came up.

It was bad stuff—jungle country, deep thick luxuriant grass whose roots still held last night's dew. Unless they were very lucky, they couldn't hope to find the ball. After a few minutes' search Goldfinger and his caddie drifted away still wider to where the rough thinned out into isolated tufts. That's good, thought Bond. That wasn't anything like the line. Suddenly he trod on something. Hell and damnation. Should he stamp in it? He shrugged his shoulders, bent down and gently uncovered the ball so as not to improve the lie. Yes, it was a Dunlop 65. 'Here you are,' he called grudgingly. 'Oh no, sorry. You play with a Number One, don't you?'

'Yes,' came back Goldfinger's voice impatiently.

'Well, this is a Number Seven.' Bond picked it up and walked over to Goldfinger.

Goldfinger gave the ball a cursory glance. He said, 'Not mine,' and went on poking among the tufts with the head of his driver.

It was a good ball, unmarked and almost new. Bond put it in his pocket and went back to his search. He glanced at his watch. The statutory five minutes was almost up. Another half-minute and by God he was going to claim the hole. Strict rules of golf, Goldfinger had stipulated. All right my friend, you shall have them!

Goldfinger was casting back towards Bond, diligently prodding and shuffling through the grass.

Bond said, 'Nearly time, I'm afraid.'

Goldfinger grunted. He started to say something when there came a cry from his caddie, 'Here you are, sir. Number One Dunlop.'

Bond followed Goldfinger over to where the caddie stood on a small plateau of higher ground. He was pointing down. Bond bent and inspected the ball. Yes, an almost new Dunlop One and in an astonishingly good lie. It was miraculous—more than miraculous. Bond stared hard from Goldfinger to his caddie. 'Must have had the hell of a lucky kick,' he said mildly.

The caddie shrugged his shoulders. Goldfinger's eyes were calm, untroubled. 'So it would seem.' He turned to his caddie. 'I think we can get a spoon to that one, Foulks.'

Bond walked thoughtfully away and then turned to watch the shot. It was one of Goldfinger's best. It soared over a far shoulder of rough towards the green. Might just have caught the bunker on the right.

Bond walked on to where Hawker, a long blade of grass dangling from his wry lips, was standing on the fairway watching the shot finish. Bond smiled bitterly at him. He said in a controlled voice, 'Is my good friend in the bunker, or is the bastard on the green?'

'Green, sir,' said Hawker unemotionally.

Bond went up to his ball. Now things had got tough again. Once more he was fighting for a half after having a certain win in his pocket. He glanced towards the pin, gauging the distance. This was a tricky one. He said, 'Five or six?'

'The six should do it, sir. Nice firm shot.' Hawker handed him the club.

Now then, clear your mind. Keep it slow and deliberate. It's an easy shot. Just punch it so that it's got plenty of zip to get up the bank and on to the green. Stand still and head down. Click! The ball, hit with a slightly closed face, went off on just the medium trajectory Bond had wanted. It pitched below the bank. It was perfect! No, damn it. It had hit the bank with its second bounce, stopped dead, hesitated and then rolled back and down again. Hell's bells! Was it Hagen who said, 'You drive for show, but you putt for dough'? Getting dead from below that bank was one of the most difficult putts on the course. Bond reached for his cigarettes and lit one, already preparing his mind for the next crucial shot to save the hole—so long as that bastard Goldfinger didn't hole his from thirty feet!

Hawker walked along by his side. Bond said, 'Miracle finding that ball.'

'It wasn't his ball, sir.' Hawker was stating a fact.

'What do you mean?' Bond's voice was tense.

'Money passed, sir. White, probably a fiver. Foulks must have dropped that ball down his trouser leg.'

'Hawker!' Bond stopped in his tracks. He looked round. Goldfinger and his caddie were fifty yards away, walking slowly towards the green. Bond said fiercely, 'Do you swear to that? How can you be sure?'

Hawker gave a half-ashamed, lop-sided grin. But there was a

crafty belligerence in his eye. 'Because his ball was lying under my
bag of clubs, sir.' When he saw Bond's open-mouthed expression
he added apologetically, 'Sorry, sir. Had to do it after what he's
been doing to you. Wouldn't have mentioned it, but I had to let
you know he's fixed you again.'

Bond had to laugh. He said admiringly, 'Well, you are a card,
Hawker. So you were going to win the match for me all on your
own!' He added bitterly, 'But, by God, that man's the flaming limit.
I've got to get him. I've simply got to. Now let's think!' They walked
slowly on.

Bond's left hand was in his trousers pocket, absent-mindedly
fingering the ball he had picked up in the rough. Suddenly the message went to his brain. Got it! He came close to Hawker. He glanced
across at the others. Goldfinger had stopped. His back was to Bond
and he was taking the putter out of his bag. Bond nudged Hawker.
'Here, take this.' He slipped the ball into the gnarled hand. Bond
said softly, urgently, 'Be certain you take the flag. When you pick
up the balls from the green, whichever way the hole has gone, give
Goldfinger this one. Right?'

Hawker walked stolidly forward. His face was expressionless. 'Got
it, sir,' he said in his normal voice. 'Will you take the putter for
this one?'

'Yes.' Bond walked up to his ball. 'Give me a line, would you?'

Hawker walked up on to the green. He stood sideways to the
line of the putt and then stalked round to behind the flag and
crouched. He got up. 'Inch outside the right lip, sir. Firm putt. Flag,
sir?'

'No. Leave it in, would you.'

Hawker stood away. Goldfinger was standing by his ball on the
right of the green. His caddie had stopped at the bottom of the
slope. Bond bent to the putt. Come on, Calamity Jane! This one
has got to go dead or I'll put you across my knee. Stand still. Club
head straight back on the line and follow through towards the hole.
Give it a chance. Now! The ball, hit firmly in the middle of the
club, had run up the bank and was on its way to the hole. But too
hard, damn it! Hit the stick! Obediently the ball curved in, rapped
the stick hard and bounced back three inches—dead as a doornail!

Bond let out a deep sigh and picked up his discarded cigarette.
He looked over at Goldfinger. Now then, you bastard. Sweat that
one out. And by God if you hole it! But Goldfinger couldn't afford

to try. He stopped two feet short. 'All right, all right,' said Bond generously. 'All square and one to go.' It was vital that Hawker should pick up the balls. If he had made Goldfinger hole the short putt it would have been Goldfinger who would have picked the ball out of the hole. Anyway, Bond didn't want Goldfinger to miss that putt. That wasn't part of the plan.

Hawker bent down and picked up the balls. He rolled one towards Bond and handed the other to Goldfinger. They walked off the green, Goldfinger leading as usual. Bond noticed Hawker's hand go to his pocket. Now, so long as Goldfinger didn't notice anything on the tee!

But, with all square and one to go, you don't scrutinize your ball. Your motions are more or less automatic. You are thinking of how to place your drive, of whether to go for the green with the second or play to the apron, of the strength of the wind—of the vital figure four that must somehow be achieved to win or at least to halve.

Considering that Bond could hardly wait for Goldfinger to follow him and hit, just once, that treacherous Dunlop Number Seven that looked so very like a Number One, Bond's own drive down the four hundred and fifty yard eighteenth was praiseworthy. If he wanted to, he could now reach the green—if he wanted to!

Now Goldfinger was on the tee. Now he had bent down. The ball was on the peg, its lying face turned up at him. But Goldfinger had straightened, had stood back, was taking his two deliberate practice swings. He stepped up to the ball, cautiously, deliberately. Stood over it, waggled, focusing the ball minutely. Surely he would see! Surely he would stop and bend down at the last minute to inspect the ball! Would the waggle never end? But now the club head was going back, coming down, the left knee bent correctly in towards the ball, the left arm straight as a ramrod. Crack! The ball sailed off, a beautiful drive, as good as Goldfinger had hit, straight down the fairway.

Bond's heart sang. Got you, you bastard! Got you! Blithely Bond stepped down from the tee and strolled off down the fairway planning the next steps which could now be as eccentric, as fiendish as he wished. Goldfinger was beaten already—hoist with his own petard! Now to roast him, slowly, exquisitely.

Bond had no compunction. Goldfinger had cheated him twice and got away with it. But for his cheats at the Virgin and the seventeenth, not to mention his improved lie at the third and the various times he had tried to put Bond off, Goldfinger would have been beaten by now. If it needed one cheat by Bond to rectify the score-sheet that was only poetic justice. And besides, there was more to this than a game of golf. It was Bond's duty to win. By his reading of Goldfinger he *had* to win. If he was beaten, the score between the two men would have been equalized. If he won the match, as he now had, he would be two up on Goldfinger— an intolerable state of affairs, Bond guessed, to a man who saw himself as all powerful. This man Bond, Goldfinger would say to himself, has something. He has qualities I can use. He is a tough adventurer with plenty of tricks up his sleeve. This is the sort of man I need for—for what? Bond didn't know. Perhaps there would be nothing for him. Perhaps his reading of Goldfinger was wrong, but there was certainly no other way of creeping up on the man.

Goldfinger cautiously took out his spoon for the longish second over cross-bunkers to the narrow entrance to the green. He made one more practice swing than usual and then hit exactly the right, controlled shot up to the apron. A certain five, probably a four. Much good would it do him!

Bond, after a great show of taking pains, brought his hands down well ahead of the club and smothered his number three iron so that the topped ball barely scrambled over the cross-bunkers. He then wedged the ball on to the green twenty feet past the pin. He was where he wanted to be—enough of a threat to make Goldfinger savour the sweet smell of victory, enough to make Goldfinger really sweat to get his four.

And now Goldfinger really was sweating. There was a savage grin of concentration and greed as he bent to the long putt up the bank and down to the hole. Not too hard, not too soft. Bond could read every anxious thought that would be running through the man's mind. Goldfinger straightened up again, walked deliberately across the green to behind the flag to verify his line. He walked slowly back beside his line, brushing away—carefully, with the back of his hand—a wisp or two of grass, a speck of top-dressing. He bent again and made one or two practice swings and then stood to the putt, the veins standing out on his temples, the cleft of concentration deep between his eyes.

Goldfinger hit the putt and followed through on the line. It was a beautiful putt that stopped six inches past the pin. Now Goldfinger would be sure that unless Bond sank his difficult twenty-footer, the match was his!

Bond went through a long rigmarole of sizing up his putt. He took his time, letting the suspense gather like a thunder cloud round the long shadows on the livid, fateful green.

'Flag out, please. I'm going to sink this one.' Bond charged the words with a deadly certitude, while debating whether to miss the hole to the right or the left or leave it short. He bent to the putt and missed the hole well on the right.

'Missed it, by God!' Bond put bitterness and rage into his voice. He walked over to the hole and picked up the two balls, keeping them in full view.

Goldfinger came up. His face was glistening with triumph. 'Well, thanks for the game. Seems I was just too good for you after all.'

'You're a good nine handicap,' said Bond with just sufficient sourness. He glanced at the balls in his hand to pick out Goldfinger's and hand it to him. He gave a start of surprise. 'Hullo!' He looked sharply at Goldfinger. 'You play a Number One Dunlop, don't you?'

'Yes, of course.' A sixth sense of disaster wiped the triumph off Goldfinger's face. 'What is it? What's the matter?'

'Well,' said Bond apologetically. "Fraid you've been playing with the wrong ball. Here's my Penfold Hearts and this is a Number Seven Dunlop.' He handed both balls to Goldfinger. Goldfinger tore them off his palm and examined them feverishly.

Slowly the colour flooded over Goldfinger's face. He stood, his mouth working, looking from the balls to Bond and back to the balls.

Bond said softly, 'Too bad we were playing to the rules. Afraid that means you lose the hole. And, of course, the match.' Bond's eyes observed Goldfinger impassively.

'But, but . . .'

This was what Bond had been looking forward to—the cup dashed from the lips. He stood and waited, saying nothing.

Rage suddenly burst Goldfinger's usually relaxed face like a bomb. 'It was a Dunlop Seven you found in the rough. It was your caddie that gave me this ball. On the seventeenth green. He gave me the wrong ball on purpose, the damned che—'

'Here, steady on,' said Bond mildly. 'You'll get a slander action on your hands if you aren't careful. Hawker, did you give Mr Goldfinger the wrong ball by mistake or anything?'

'No, sir.' Hawker's face was stolid. He said indifferently, 'If you want my opinion, sir, the mistake may have been made at the seventeenth when the gentleman found his ball pretty far off the line we'd all marked it on. A Seven looks very much like a One. I'd say that's what happened, sir. It would have been a miracle for the gentleman's ball to have ended up as wide as where it was found.'

'Tommy rot!' Goldfinger gave a snort of disgust. He turned angrily on Bond. 'You saw that was a Number One my caddie found.'

Bond shook his head doubtfully. 'I didn't really look closely, I'm afraid. However,' Bond's voice became brisk, businesslike, 'it's really the job of the player to make certain he's using the right ball, isn't it? I can't see that anyone else can be blamed if you tee the wrong ball up and play three shots with it. Anyway,' he started walking off the green, 'many thanks for the match. We must have it again one day.'

Goldfinger, lit with glory by the setting sun, but with a long black shadow tied to his heels, followed Bond slowly, his eyes fixed thoughtfully on Bond's back.   IAN FLEMING, *Goldfinger*, 1959

These scenes from Walker Percy's excellent novel have an extraordinary intensity of feeling and physical reality.

Undoubtedly something was happening to him.

It began again the next day when he sliced out-of-bounds and was stooping through the barbed-wire fence to find his ball. For the first time in his life he knew that something of immense importance was going to happen to him and that he would soon find out what it was. Ed Cupp was holding the top strand high so he could crawl through, higher than he needed to, to make up for his, Ed Cupp's, not following him into the woods to help him find the ball. To prove his good intentions, Ed Cupp pulled the wire so hard that it stretched as tight as a guitar string and creaked and popped against the fence posts.

As he stopped and in the instant of crossing the wire, head lowered, eyes lightly bulging and focused on the wet speckled leaves marinating and funky-smelling in the sunlight, he became aware that he was doing an odd thing with his three-iron. He was holding it in his left hand, fending against the undergrowth with his right and turning his body into the vines and briars which grew in the fence so that they snapped against his body. Then, even as he was climbing through, he had shifted his grip on the iron so that the club head was tucked high under his right arm, shaft resting on forearm, right hand holding the shaft steady—as one might carry a shotgun.

He did not at first know why he did this. Then he did know why.

Now he was standing perfectly still in a glade in a pine forest holding the three-iron, a good fifty feet out-of-bounds and not looking for the ball. It was only after standing so for perhaps thirty seconds, perhaps two minutes, that he made the discovery. The discovery was that he did not care that he had sliced out-of-bounds.

A few minutes earlier he had cared. As his drive curved for the woods, the other players watched in silence. There was a mild perfunctory embarrassment, a clucking of tongues, a clearing of throats in a feigned but amiable sympathy.

Lewis Peckham, the pro, a grave and hopeful man, said: 'It could have caught that limb and dropped fair.'

Jimmy Rogers, a man from Atlanta, who had joined the foursome to make it an unwieldy fivesome, said: 'For a six-handicapper and a Wall Street lawyer, Billy is either nervous about his daughter's wedding or else he's taking it easy on his future in-laws.'

He hit another ball and it too sliced out-of-bounds.

The other four golfers gazed at the dark woods in respectful silence and expectation as if they were waiting for some rule of propriety to prevail and to return the ball to the fairway.

As he leaned over to press the tee into the soft rain-soaked turf, he felt the blood rush to his face. Jimmy Rogers had gotten on his nerves. Was it because Jimmy Rogers had messed up the foursome or because Jimmy Rogers had called him Billy? How did Jimmy Rogers know his handicap?

After teeing up the third ball and as he measured the driver and felt his weight shift from one foot to the other, he was wondering absentmindedly: What if I slice out-of-bounds again, what then? Is a game so designed that there is always a chance that one can so badly transgress its limits and bounds, fall victim to its hazards, that disgrace is always possible, and that it is the public avoidance of disgrace that gives one a pleasant sense of license and justification?

He sliced again but not out-of-bounds, having allowed for the slice by aiming his stance toward the left rough.

He said: 'I'm picking up. It's the eighteenth anyhow. I'll see you in the clubhouse.'

The slice, which had become worrisome lately, had gotten worse. He had come to see it as an emblem of his life, a small failure at living, a minor deceit, perhaps even a sin. One cringes past the ball, hands mushing through ahead of the club in a show of form, rather than snapping the club head through in an act of faith. Unlike sin in life, retribution is instantaneous. The ball, one's very self launched into its little life, gives offense from the very outset, is judged, condemned, and sent screaming away and, banished from the pleasant licit fairways and the sunny irenic greens, goes wrong and ever wronger, past the rough, past even the barbed-wire fence, and into the bark fens and thickets and briars of out-of-bounds. One is punished on the spot.

Earlier he had seen a bird, undoubtedly some kind of a hawk, fly across the fairway straight as an arrow and with astonishing swiftness, across a ridge covered by scarlet and gold trees, then fold its wings and drop like a stone into the woods. It reminded him of something but before he could think what it was, sparks flew forward at the corner of his eye. He decided with interest that something was happening to him, perhaps a breakdown, perhaps a stroke. When his turn came to putt and he stooped over the ball, he looked at the hole some twenty feet away and at Lewis Peckham, who was tending the pin and who was looking not at him or the hole but in a small exquisite courtesy allowing his eyes to go unfocused and gaze at a middle distance. The green broke to the right. He did not know whether he was going to hit the ball five feet or fifty feet. It was as if the game had fallen away from him and he was trying to play it from a great height. He felt like a clown on stilts. Lewis Peckham cleared his throat and now Lewis was looking at him and his eyes were veiled and ironic (as if he not only knew that something was happening to him but even knew what it was!) and he putted. The ball curved in a smooth flat parabola and sank with a plop.

It was a good putt. His muscles remembered. When the putt sank, the golfers nodded briefly, signifying approval and a kind of relief that he was back on his game. Or was it a relief that they could play a game at all, obey its rules, observe its etiquette and the small rites of settling in for a drive and lining up a putt? He was of two minds, playing golf and at the same time wondering with no more than a moderate curiosity what was happening to

him. Were they of two minds also? Was there an unspoken understanding between all of them that what they were doing, knocking little balls around a mountain meadow while the fitful wind
bustled about high above them, was after all preposterous but that
they had all assented to it and were doing it nevertheless and
because, after all, why not? One might as well do one thing as
another.

But the hawk was not of two minds. Single-mindedly it darted
through the mountain air and dove into the woods. Its change of
direction from level flight to drop was fabled. That is, it made him
think of times when people told him fabulous things and he believed
them. Perhaps a Negro had told him once that this kind of hawk
is the only bird in the world that can—can what? He remembered.
He remembered everything today. The hawk, the Negro said, could
fly full speed and straight into the hole of a hollow tree and brake
to a stop inside. He, the Negro, had seen one do it. It was possible to believe that the hawk could do just such a fabled single-
minded thing. WALKER PERCY, *The Second Coming*, 1981

MCAUSLAN IN THE ROUGH
My tough granny—the Presbyterian MacDonald one, not the pagan
one from Islay—taught me about golf when I was very young. Her
instruction was entirely different from that imparted by my father,
who was a scratch player, gold medallist and all, with a swing like
de Vicenzo; he showed me how to make shots, and place my feet,
and keep calm in the face of an eighteen-inch putt on a downhill
green with the wind in my face and the match hanging on it. But
my granny taught me something much more mysterious.

Her attitude to the game was much like her attitude to religion;
you achieved grace by sticking exactly to the letter of the law, by
never giving up, and by occasional prayer. You replaced your divots, you carried your own clubs, and you treated your opponent
as if he was a Campbell, and an armed one at that. I can see her
now, advanced in years, with her white hair clustered under her black
bonnet, and the wind whipping the long skirt round her ankles,
lashing her drives into the gale; if they landed on the fairway she
said 'Aye', and if they finished in the rough she said 'Tach!' Nothing
more. And however unplayable her lie, she would hammer away
with her niblick until that ball was out of trouble, and half Perthshire

with it. If it took her fifteen strokes, no matter; she would tot them
up grimly when the putts were down, remark, 'This and better
may do, this and waur will never do', and stride off to the next
tee, gripping her driver like a battle-axe.

As an opponent she was terrifying, not only because she played
well, but because she made you aware that this was a personal duel
in which she intended to grind you into the turf without pity;
if she was six up at the seventeenth she would still attack that
last hole as if life depended on it. At first I hated playing with her,
but gradually I learned to meet her with something of her own
spirit, and if I could never achieve the killer instinct which she
possessed, at least I discovered satisfaction in winning, and did so
without embarrassment.

As a partner she was beyond price. Strangely enough, when we
played as a team, we developed a comradeship closer than I ever
felt for any other player; we once even held our own with my
father and uncle, who together could have given a little trouble to
any golfers anywhere. Even conceding a stroke a hole they were
immeasurably better than an aged woman and an erratic small boy,
but she was their mother and let them know it; the very way she
swung her brassie was a wordless reminder of the second com-
mandment, and by their indulgence, her iron will, and enormous
luck, we came all square to the eighteenth tee.

Counting our stroke, we were both reasonably close to the green
in two, and my granny, crouching like a bombazine vulture with
her mashie-niblick, put our ball about ten feet from the pin. My
father, after thinking and clicking his tongue, took his number three
and from a nasty lie played a beautiful rolling run-up to within a
foot of the hole—a real old Fife professional's shot.

I looked at the putt and trembled. 'Dand,' said my grandmother. 'Never up, never in.'

So I gulped, prayed, and went straight for the back of the cup.
I hit it, too, the ball jumped, teetered, and went in. My father and
uncle applauded, granny said 'Aye', and my uncle stooped to his
ball, remarking, 'Halved hole and match, eh?'

'No such thing,' said granny, looking like the Three Fates. 'Take
your putt.'

Nowadays, of course, putts within six inches or so are frequently conceded, as being unmissable. Not with my grandmother; she
would have stood over Arnold Palmer if he had been on the lip of

the hole. So my uncle sighed, smiled, took his putter, played—and missed. His putter went into the nearest bunker, my father walked to the edge of the green, humming to himself, and my grandmother sniffed and told me curtly to pick up my bag and mind where I was putting my feet on the green.

As we walked back to the clubhouse, she grimly silent as usual, myself exulting, while the post-mortem of father and uncle floated out of the dusk behind us, she made one of her rare observations. 'A game,' she said, 'is not lost till it's won. Especially with your Uncle Hugh. He is—' and here her face assumed the stern resignation of a materfamilias who has learned that one of the family has fled to Australia pursued by creditors, '—a *trifling* man. Are your feet wet? Aye, well, they won't stay dry long if you drag them through the grass like that.'

And never a word did she say about my brilliant putt, but back in the clubhouse she had the professional show her all the three irons he had, chose one, beat him down from seventeen and six to eleven shillings, handed it to me, and told my father to pay for it. 'The boy needs a three iron,' she said. And to me: 'Mind you take care of it.' I have taken care of it.

GEORGE MACDONALD FRASER, McAuslan in the Rough, 1974

OUR GOLF BALLS

Imagine the spray of atoms jolted out of the golf ball's very being by the impact of the driver. In a close-up we see the ball's upper half spilling over the sharp edge of the club, while the lower half begins the incredible flight.

You believe the golf ball achieves maximum velocity at impact, but in fact this occurs several yards from the tee, and at the moment of greatest speed, when at its finest as a golf ball, it is flat, squashed against the air, an odd oblong resembling in my excellent photographs a forward-leaning ghost stumbling through space.

Exactly how fast is our golf ball traveling? We all would of course like to know, but even our most careful estimates smack of the wayward melancholy of oafs speculating in a pasture: Is there a face in the sun? Is a beautiful thing always beautiful or does it give over? In Heaven where will we pee—in the clouds? Science has turned its back on the question of the golf ball's speed, thickheadedly assuming accelerologists settled the matter in the days of the immortal Tommy Armour! But a safe and considered estimate of the golf ball's maximum velocity is six hundred and fifty miles per hour.

Carefully following its flight, we find it not only stumbles, it goes end over end: a ludicrous, not to say dizzying, moment unequaled in human experience—as far as we live to tell. During this moment the golf ball would, if it could, lose consciousness. I have paused to watch men and women in similar moments of excruciation; I'm sure you've watched too; we shall be watched when our time comes. Our shrinks slyly assure us we blot out the ultimately insufferable, but there are those who seem unfazed during great moments: our true heroes! They are capable of grandeur because they remain miraculously inviolate through those occasions when Heaven and Hell are one, when all things swirl off into light, and great Nothing roars like the sea. But they are shamming. They are no braver than our golf ball silently squealing its terror through space, and I have proof—a photo in which a racer is blurrily seen at the moment he gets a speed record: he hunkers in the cockpit of his jet-powered, fifty-thousand-pound, four-wheeled bullet, his face invisible beneath the black visor of his helmet. In this priceless photograph we penetrate distance, the blur of speed, the black visor, and see that our hero's moment of ultimate triumph is also his moment of ultimate shame: he's driving nine hundred miles per hour with his eyes shut!

We left our golf ball going through space end over end. Sun—earth—sun. . . . Then the silent descent, the deep breath of sanity.

The ball lands, hops along, stops, and can be seen, an object no bigger than the end of my thumb, from a distance of three hundred yards. How can this be?

After a tee shot the ball quite easily, though miraculously, sprawls! It is one of those ominous, necessarily shunned but daily miracles which crowd our days, making us yearn for boredom.

It's again itself—a tidy, round golf ball—when the player, clutching a club with a shiny steel head, comes tromping up. The player glances about, as if to see if anyone is watching, then drawing back, he begins viciously hacking the ball. The golf ball becomes an intimate of the two iron, the three, the inexorable five iron.

The golf ball lies on the green which is a sloping sky. Little taps send it rolling up and down—but in the wrong direction! It is being aimed toward a hole in the sky, and each time it passes it hears the cold suck of wind down the hole. Other balls slide up and down and our golf ball looks off, as only balls can, as one by one the others slip their grip and are gone.

We stand atop tall buildings and look at the cars and tiny people below. One hundred stories is as high as we dare go; higher, and we lose interest in looking down and, staring off, expect planes, birds and soon, vaguely, even things of miraculous bearing. We also stand, the tribe of Man, at great cracks and look at rivers writhing below. But always we stand firmly on the earth. We may throb with the bigness and looseness of things—'Ah! A wide vista!'— and we may suffer somewhat. But the earth is under our feet and we can always see the bottom of things.

Our golf ball, though, slides down a green sky that slants to a hole, and there's no wide vista: imagine looking into a hole, lit by tiny pinpoint light bulbs, four inches wide and a mile deep.

If you have succeeded in this, next imagine you are a golf ball. (Densely round. Squint. Clench your teeth.) Ready?

Fall.   JERRY BUMPUS, *Things in Place*, 1975

❖

GOLF DREAMS

They steal upon the sleeping mind while winter steals upon the landscape, sealing the inviting cups beneath sheets of ice, cloaking the contours of the fairway in snow.

I am standing on a well-grassed tee with my customary summer foursome, whose visages yet have something shifting and elusive about them. I am getting set to drive; the fairway before me is a slight dog-leg right, very tightly lined with trees, mostly conifers. As I waggle and lift my head to survey once more the intended line of flight, further complications have been imposed: the air above the fairway has been interwoven with the vines and wooden cross-pieces of an arbor, presumably grape, and the land seems to drop away no longer with a natural slope but in nicely hedged terraces. Nevertheless, I accept the multiplying difficulties calmly, and try to allow for them in my swing, which is intently contemplated but never achieved, for I awake with the club at its apogee, waiting for my left side to pull it through and to send the ball toward that bluish speck of openness beyond the vines, between the all but merged forests.

It is a feature of dream golf that the shot never decreases in difficulty but instead from instant to instant melts, as it were, into deeper hardship. A ball, for instance, lying at what the dreaming golfer gauges to be a 7-iron distance from the green, has become, while he glanced away, cylindrical in shape—a roll of coins in a paper wrapper, or a plastic bottle of pills. Nevertheless, he swings, and as he swings he realizes that the club in his hands bears a rubber tip, a little red-rubber tab the color of a crutch tip, but limp. The rubber flips negligibly across the cylindrical 'ball,' which meanwhile appears to be sinking into a small trough having to do, no doubt, with the sprinkler system. Yet, most oddly, the dreamer surrenders not a particle of hope of making the shot. In this instance, indeed, I seem to recall making, on my second or third swing, crisp contact, and striding in the direction of the presumed flight with a springy, expectant sensation.

After all, are these nightmares any worse than the 'real' drive that skips off the toe of the club, strikes the prism-shaped tee marker, and is swallowed by weeds some twenty yards *behind* the horrified driver? Or the magical impotence of an utter whiff? Or the bizarre physical comedy of a soaring slice that strikes the one telephone wire strung across three hundred acres? The golfer is so habituated to humiliation that his dreaming mind never offers any protest of implausibility. Whereas dream life, we are told, is a therapeutic caricature, seamy side out, of real life, dream golf is simply golf played on another course. We chip from glass tables onto moving stairways; we swing in a straitjacket, through masses of cobweb, and awake not with any sense of unjust hazard but only with a regret that the round can never be completed, and that one of our phantasmal companions has kept the scorecard in his pocket.

Even the fair companion sleeping beside us has had a golf dream, with a feminist slant. An ardent beginner, she says, 'I was playing with these men, I don't know who they were, and they kept using woods when we were on the green, so of course the balls would fly miles away, and then they had to hit all the way back. I thought to myself, *They aren't using the right club*, and I took my putter out and, of course, I kept *beating* them!'

'Didn't they see what you were doing, and adjust their strokes accordingly?'

'No, they didn't seem to *get* it, and I wasn't going to tell them. I kept *winning*, and it was *wonderful*,' she insists.

We gaze at each other across the white pillows, in the morning
light filtered through icicles, and realize we were only dreaming.
Our common green hunger begins to gnaw afresh, insatiable.

JOHN UPDIKE, *New Yorker*, 19 February 1979

THE GOLF COURSE OWNER

He sits by the little clubhouse, in a golf cart, wearing black. He is
Greek. Where, after all these years in America, does he buy black
clothes? His hat is black. His shirt is black. His eyes, though a bit
rheumy with age now, are black, as are his shoes and their laces.
Small black points exist in his face, like scattered punctuation. His
smile is wonderful, an enfolding of the world as his hand enfolds
yours. Many little gray teeth, all his, they must be: something of
the ancient marriage of tragedy and comedy in that smile.

How ancient is he? He has been sitting here since one learned
golf twenty years ago. In those years, it was his son who manned
the tractor with his gang of mowers, going up and down the fairways methodically as a lover's caresses. Now, it is his grandson.
Once, in Homeric times, it must have been he, Harílaos, who
manned the tractors. But times so epic are hard to imagine.

The first, second, third, and ninth holes can be seen from where
he sits, and the fourth tee, where many a man has been tempted
by the broad downhill leftward dog-leg to hook into the marsh.
The ridge holds its writhing occupants in profile, a frieze against
the sky, before they mourn their shots, pocket their tees, and drag
their carts down into the underworld. The fifth, sixth, seventh, and
eighth holes are entirely out of sight, but the men in their bright
slacks eventually return, advancing down the ninth fairway like a
thinned army pulling its own chariots. Their odyssey ends in a ritual exchange with the owner, who has that essential capacity myth-
ic characters have for waiting, waiting decades if need be, for the
foreordained moment in the adventure to arrive.

Q: How goes it, Harry, how goes it?
A: Not so good, John, not so good.
Q: Lovely day out there.
A: [*nods*]

Weather and health are discussed but never, oddly, golf. What
does he know about golf? Among the mysteries that radiate as he
in his black clothes soaks up the sun are:

Q: How did he come to acquire this frivolously utilized acreage?
Q: Does it turn a profit?
Q: What *is* this Greek genius for acquisition?

JOHN UPDIKE, *Ontario Review*

Although not a regular player, Julian Barnes says he knows enough about the game to savour the pleasure of the well-hit shot, and Uncle Leslie's delight in the Old Green Heaven will be recognized by many—even if they express it less eccentrically.

Other people assumed it must be a strain, looking back over ninety years. Tunnel vision, they guessed; straw vision. It wasn't like that. Sometimes the past was shot with a hand-held camera; sometimes it reared monumentally inside a proscenium arch with moulded plaster swags and floppy curtains; sometimes it eased along, a love story from the silent era, pleasing, out of focus and wholly implausible. And sometimes there was only a succession of stills to be borrowed from the memory.

The Incident with Uncle Leslie—the very first Incident of her life—came in a series of magic lantern slides. A sepia morality; the lovable villain even had a moustache. She had been seven at the time; it was Christmas; Uncle Leslie was her favourite uncle. Slide 1 showed him bending down from his enormous height to hand over a present. Hyacinths, he whispered, giving her a biscuit-coloured pot surmounted by a mitre of brown paper. Put them in the airing cupboard and wait until the Spring. She wanted to see them now. Oh, they wouldn't be up yet. How could he be sure? Later, in secret, Leslie unscrewed a corner of the brown-paper wrapping and let her peer in. Surprise! They were up already. Four slim ochre points, about half an inch long. Uncle Leslie emitted the reluctant chuckle of an adult suddenly impressed by a child's greater knowledge. Still, he explained, this was all the more reason why she shouldn't look at them again until the Spring; any more light could cause them to outgrow their strength.

She put the hyacinths in the airing cupboard and waited for them to grow. She thought about them a lot, and wondered what a hyacinth looked like. Time for Slide 2. In late January she went to the bathroom with a torch, turned off the light, took down the pot, unscrewed a tiny viewing-hole, aimed the torch and quickly looked inside. The four promising tips were still there, still half an inch long. At least the light she had let in at Christmas hadn't harmed them.

In late February she looked again; but obviously the growing season hadn't started yet. Three weeks later Uncle Leslie called by on his way to play golf. Over lunch he turned to her conspiratorially and asked, 'Well, little Jeanie, are the hyacinths hyacinth Christmas?'

'You told me not to look.'

'So I did. So I did.'

She looked again at the end of March, then—Slides 5 to 10—on April the second, fifth, eighth, ninth, tenth and eleventh. On the twelfth her mother agreed to a closer examination of the pot. They laid yesterday's Daily Express on the kitchen table, and carefully unwrapped the brown paper. The four ochre sprouts had not advanced at all. Mrs Serjeant looked uneasy.

'I think we'd better throw them out, Jean.' Adults were always throwing things out. That was clearly one of the big differences. Children liked keeping things.

'Maybe the roots are growing.' Jean started easing away at the peaty earth packed tight against the tips.

'I shouldn't do that.' said Mother. But it was too late. One after the other, Jean dug out four upturned wooden golf tees.

Strangely, the Incident didn't make her lose faith in Uncle Leslie. Instead, she lost faith in hyacinths.

Looking back, Jean assumed that she must have had friends as a child; but she couldn't recall that special confidante with the wonky grin, or the playground games with skipping-rope and acorns, or the secret messages passed along ink-stained desks at a village school with a daunting stone inscription above its door. Perhaps she had had all these things: perhaps not. In retrospect, Uncle Leslie had been friends enough. He had crinkly hair which he kept well Brylcreemed, and a dark blue blazer with a regimental badge on the breast pocket. He knew how to make wine glasses out of sweet papers, and whenever he went to the golf club he always called it 'popping down the Old Green Heaven'. Uncle Leslie was the sort of man she would marry.

Shortly after the hyacinth Incident, he began taking her down the Old Green Heaven. When they arrived he would sit her on a mildewed bench near the car-park and instruct her with mock severity to guard his clubs.

'Just going to wash behind the old ear-pieces.'

Twenty minutes later they would set off towards the first tee. Uncle Leslie carrying his clubs and smelling of beer, Jean with the sand-iron over her shoulder. This was a good-luck ploy devised by Leslie: as long as Jeanie was carrying the sand-iron in readiness, the lightning would be diverted and he would be kept out of the bunker.

'Don't let the club-head drop,' he would say, 'or there'll be more sand flying than on a windy day in the Gobi desert.' And she would shoulder the club correctly, like a rifle. Once, feeling tired at the uphill fifteenth, she had trailed it behind her off the tee, and Uncle Leslie's second shot squirted straight into a bunker fifteen yards away.

'Now look what you've done,' he said; though he seemed almost as pleased as he was cross. 'Have to buy me one at the nineteenth for that.'

Uncle Leslie often talked to her in a funny code she pretended to understand. Everyone knew there were only eighteen holes on a golf course, and that she didn't have any money, but she nodded as if she was always buying people one—one what?—at the nineteenth. When she grew up, someone would explain the code to her; in the meantime she felt quite happy not knowing. And there were bits she understood already. If the ball swerved disobediently off into the woods, Leslie would sometimes mutter, 'One for the hyacinths' —the only reference he ever made to his Christmas present.

But mostly his remarks were beyond her. They marched pur-posefully down the fairway, he with his bagful of quietly clanking hickory, she sloping arms with the sand-iron. Jean was not allowed to talk: Uncle Leslie had explained that chatter put him off think-ing about his next shot. He, on the other hand, was allowed to speak; and as they strode towards that distant white glint which sometimes turned out to be a sweet paper, he would occasionally stop, bend down and whisper to her the secrets of his mind. At the fifth he told her that tomatoes were the cause of cancer, and that the sun would never set on the Empire; at the tenth she learned that bombers were the future, and that old Musso might be an Eytie but he knew which way the paper folded. Once they had stopped on the short twelfth (an unprecedented act on a par three) while Leslie gravely explained, 'Besides, your *Jew* doesn't really *enjoy* golf.'

Then they had continued towards the bunker on the left of the green, with Jean repeating to herself this suddenly awarded truth.

She liked going down the Old Green Heaven; you never knew quite what would happen. Once, after Uncle Leslie had washed behind his ears more thoroughly than usual, he had crackled off into the deep rough alongside the fourth. She was made to turn her back, but couldn't avoid hearing a prolonged splashing noise of remarkable volume and implications. She had peered under a raised elbow (it didn't count as looking), and seen steam rising amid the waist-high bracken.

Next there was Leslie's trick. Between the ninth green and the tenth tee, surrounded by newly planted silver birches, was a little wooden hut like a rustic bird-box. Here, if the wind was in the right direction, Uncle Leslie would sometimes do his trick. From the breast pocket of his tweed jacket with the leather elbows he would take a cigarette, lay it on his knee, pass his hands over it like a magician, put it in his mouth, give Jeanie a slow wink, and strike a match. She would sit beside him trying to hold her breath, trying not to be a shufflebottom. Huffers and puffers spoiled tricks, Uncle Leslie had said, and so did shufflebottoms.

After a minute or two she would ease her glance sideways, taking care not to move suddenly. The cigarette had an inch of ash on it, and Uncle Leslie was taking another puff. At the next glance, his head was tipped slightly back, and half the cigarette consisted of ash. From this point on, Uncle Leslie wouldn't look at her; instead, he would concentrate very carefully, slowly leaning back a little more with each puff he took. Finally, his head would be at right-angles to his spine, with the cigarette, now pure ash apart from the last half-inch where Leslie was holding it, rising vertically towards the roof of the giant bird-box. The trick had worked.

Then he would reach out his left hand and touch her upper arm; she would get up quietly, trying not to breathe in case she huffed and puffed the ash down Leslie's jacket with the leather elbows, and go ahead to the tenth tee. A couple of minutes later Leslie would rejoin her, smiling a little. She never asked how he did his trick; perhaps she thought he wouldn't tell her.

And then there was the screaming. This always happened in the same place, a field behind the triangle of damp, smelly beeches which pushed their way in to the dogleg fourteenth. On each occasion, Uncle Leslie had sliced his drive so badly that they had to search the least visited part of the wood, where the trunks had moss on them and the beech-nuts were thicker on the ground. The first time, they had found themselves by a stile, which was slimy to the touch though the weather had been dry for days. They climbed the stile and began hunting in the first few yards of sloping meadowland. After some rather aimless kicking and club-scuffling, Leslie had bent down and said, 'Why don't we have a good old scream?'

She smiled back at him. Having a good old scream was clearly something people did on these occasions. After all, it was very annoying not to be able to find the ball. Leslie explained further. 'When you're all screamed out you have to fall down. That's the rules.'

Then they put their heads back and screamed at the sky. Uncle Leslie deep and throaty, like a train coming out of a tunnel; Jean high and wavering, uncertain how long her breath would last. You kept your eyes open—that seemed to be an unstated rule—and stared hard up at the sky, daring it to answer your challenge. Then you took your second breath, and screamed again, more confidently, more insistently. Then again, and in the pause for each fresh breath Leslie's train noises swelled and roared; and then exhaustion arrived suddenly, and you had no scream left, and you fell to the ground. She would have fallen anyway, even if it hadn't been in the rules; fatigue raced through her body like a tidal bore.

There was a thump as Uncle Leslie flopped down a few yards away, and they stared their parallel, heaving stares up at the quiet sky. Half-way to heaven, a few small clouds shifted gently, as if reluctantly tethered; but perhaps even this movement was given them only by the panting of the two supine figures. It was clearly in the rules that you could pant as hard as you liked.

After a while, she heard Leslie cough.

'Well,' he said, 'I think I'll allow myself a free drop.' And they trailed back across the slimy stile, through the crackly beech-nuts to the angle of the fourteenth where Uncle Leslie, after looking around for spies, calmly thumbed a tee into a fairway, popped a gleaming new ball on top, and struck a brassie some two hundred yards to the green. This despite being all screamed out, thought Jean.

They went screaming only when Leslie sliced his ball very badly off the tee, which seemed to happen when the course was empty.

And they didn't do it too often, because after the first occasion Jean got the whooping cough. Getting the whooping cough hadn't qualified as an Incident, but Uncle Leslie's whip-round had. Or rather, the result of Uncle Leslie's whip-round.

She was in bed on the fourth day of her illness, occasionally giving the throaty cry of some exotic bird lost in a foreign sky, when he dropped in. He sat on her bed in his blazer with the badge, smelling a bit as if he'd been washing behind his ears, and instead of asking how she felt, murmured, 'You didn't tell them about the screaming?'

Of course she hadn't.

'Only you see, it's a secret, after all. Rather a good secret, it seems to me.'

Jean nodded. it was a remarkably good secret. But perhaps the screaming had caused the whooping cough. Her mother was always telling her to guard against over-excitement. Maybe she had over-excited her throat by screaming, and it had started whooping as a result. Uncle Leslie behaved as if he suspected things might be his fault. As she gave her panicking bird-call, he looked a little shifty.

Two days later Mrs Serjeant put Jean's winter underwear on the edge of the bed, then a thick dress, her winter coat, a scarf and a blanket. She seemed displeased but resigned.

'Come on. Uncle Leslie's had a whip-round.' Uncle Leslie's whip-round, Jean discovered, included a taxi. Her first taxi. On the way to the aerodrome she took care not to appear over-excited. At Hendon her mother stayed in the car. Jean took her father's hand, and he explained to her that the wooden parts of a De Havilland were made of spruce. Spruce was a very hard wood, he said, almost as hard as the metal parts of the aeroplane, so she was not to worry. She had not been worrying.

Sixty-minute sightseeing tour of London; departures on the hour. Among the dozen passengers were two more children wrapped up like parcels although it was only August; perhaps their uncles had had whip-rounds as well. Her father sat across the aisle and stopped her when she tried to lean past him and look out: the point of the flight, he explained, was medical, not educational. He spent the whole trip gazing at the back of the wicker seat in front of him and holding on to his knee-caps. He seemed as if he might get over-excited at any minute. When the De Havilland banked, Jean could see, beyond its chubby engines and the criss-cross of the struts, something that might be Tower Bridge. She turned to her father.

'Shh,' he said, 'I'm concentrating on getting you better.'

It was almost a year before she and Uncle Leslie went screaming again. They popped down the Old Green Heaven, of course; but somehow Leslie's driving at the dogleg fourteenth had acquired a new accuracy. When, finally, the next summer, he drew the club-head across the face of the ball and produced a high, wailing slice, the ball seemed to know exactly where it was meant to go. So did they: through the long rough, across the damp beech wood, over the slimy stile, and into the sloping meadowland. They screamed into the warm air, and thumped down on their backs. Jean found herself scanning the sky for aeroplanes. She rolled her eyes round in their sockets, and searched to the edge of her vision. No clouds, and no aeroplanes: it was as if she and Uncle Leslie had emptied the sky with their noise. Nothing but blue.

'Well,' said Leslie, 'I think I'll award myself a free drop.' They had not looked for his ball on their way through the wood, and they did not look for it on the way back either.

JULIAN BARNES, *Staring at the Sun*, 1986

GOLF POETRY

❖

No great poetry has been written about golf, but there is a fair amount of pleasant light verse. The first such work, in 1743, was Thomas Mathison's The Goff, a mock epic commemorating a match between the author, an Edinburgh lawyer, and his friend Alexander Dunning, who was the governor of Watson's Hospital. The stake was a bowl of punch.

In the match they come to the last hole all square. Pygmalion (Mathison) skies his drive; Castalio (Dunning) hits a good one, but it unfortunately strikes a sheep and kills it. The god Pan, enraged by this assault on one of his creatures, kicks the ball into a bunker and the description of the end of the game shows that absolutely nothing has changed in match play in the last 250 years.

> To free the ball the Chief now turns his mind,
> Flies to the bank where lay the orb confined,
> The ponderous club upon the ball descends,
> Involv'd in dust, th'exulting orb ascends;
> Their loud applause the pleas'd spectators raise,
> The hollow bank resounds Castalio's praise.
> A mighty blow Pygmalion then lets fall;
> Straight from th'impulsive engine starts the ball,
> Answ'ring its maker's just design, it hastes,
> And from the hole scarce twice two clubs length rests.
> Ah, what avails thy skill, since Fate decrees
> Thy conqu'ring foe to bear away the prize?
> Full fifteen clubs length from the hole he lay,
> A wide cart-road before him cross'd his way;
> The deep-cut tracks th'intrepid Chief defies,
> High o'er the road the ball triumphing flies,
> Lights on the green and scours into the hole.
> Down with it sinks depress'd Pygmalion's soul.
> Seiz'd with surprise th'affrighted hero stands;
> And feebly tips the ball with trembling hands;
> The creeping ball its want of force complains

A grassy tuft the loitering orb detains.
Surrounding crowds the victor's praise proclaim,
The echoing shore resounds Castalio's name.
For him Pygmalion must the bowl prepare,
To him must yield the honours of the war,
On fame's triumphant wings his name shall soar
Till time shall end or GOFFING be no more.

THAT CONFOUNDED BUNKER!

Last year I set my mind (after trying hard to find
   A way that was according to my liking)
That I should act the swell, and should act him very well,
   And I warrant my success was duly striking!

I heard there was a game of very ancient fame,
   And that it was the madness of the day:
 For it's 'Golf' that's Johnnies patter, and it's 'Golf' that's Pollies chatter,
   And why should I not have a try as well as they?

I confided in a friend on whose sense I could depend,
   And I asked if he would join me in the ploy.
He consented, and we started, both elated and light-hearted,
   As enamoured of the prank as any boy.

We determined to invest in an outfit of the best,
   And we spent a few good pounds upon our tools;
But I'd rather not recount—quite exactly—the amount,
   In case the thrifty-minded count us fools.

At length we reached the 'tee,' in a state of bubbling glee,
   Conspicuous by the glare of our attire;
We engaged our 'caddy' guides, and arranged to take up sides—
   In short, we stood in readiness to 'fire!'

As, with introductory cough, I was shaping to strike off,
   My friend desired that *he* should 'have the say.'
With a swoop he smote the ground, and it shook for miles around,
   And his club-head flew a thousand yards away.

His words I won't repeat—they were neither soft nor sweet—
   And his looks resembled darkest midnight thunder.
With a snort of fierce contempt he repeated his attempt,
   And his stroke might well have rent the rocks asunder.

Compared with such displays my achievements were like rays
   Of a farthing dip set up against the sun;

Though I smashed a club or two, and of balls lost not a few,
  Yet I entered, heart and soul, into the fun.

But my friend was getting low, and he felt disposed to go,
  For his stock of balls was running rather short;
His clubs were all in bits with his thundering, blundering hits,
  Yet he grumbled at the poorness of the sport!

With one final desperate shot—'twas the only one he got—
  He landed in a bunker of loose sand;
So, with a fiendish yell, to assaulting it he fell,
  And raised the dust in clouds on every hand.

For an hour I stood to see what the upshot was to be,
  And still the clouds came flying from the hole;
When I left he still was there, and I solemnly declare
  That by this time he must be a human mole.

If he does not reappear upon either hemisphere,
  There's a duty that's incumbent on the nation,
To equip and organise a Relieving Enterprise,
  And confirm it with a Royal Proclamation.   P.M.

DEDICATED TO THE DUFFER
This is the substance of our Plot—
For those who play the Perfect Shot,
There are ten thousand who do not.

For each who comes to growl and whine
Because one putt broke out of line
And left him but a Sixty-Nine,

At least ten thousand on the slate
Rise up and cheer their blessed fate
Because they got a Ninety-Eight.

For each of those who rarely sees,
Amid his run of Fours and Threes,
A Trap or Bunker—if you please—

Ten thousand Blighted Souls are found
Who daily pummel, pierce, or pound
The scourging sand-heaps underground.

Who is it pays the major fee
For rolling green and grassy tee?
Who is it, Reader?—answer me!

The scattered few in countless clubs
Who sink their putts as if in tubs,
Or eke the half a million dubs?

He may not have the Taylor Flip—
He may not know the Vardon Grip—
He may not Pivot at the Hip—

And we will say his Follow Through
Is frequently somewhat askew,
Or halting, as if clogged with Glue—

Yet, Splashers in the Wayside Brook,
To you who foozle, slice, and hook,
We dedicate This Little Book.

Not that your Style enthralls the eye
But that there are, to spring the Why,
So many more of you to Buy.

GRANTLAND RICE

KEEP YOUR EYE ON THE BALL
Boy, if the phone should ring,
    Or any one come to call,
Whisper that this is spring,
    To come again next fall.
Say I have a date on a certain tee
Where my friends the sand-traps wait in glee;
Tell them the 'Doc' has ordered me
    To keep my eye on the ball.

Boy, if they wish to know
    Where I shall haunt the scene,
Tell them to leave and go
    Out by the ancient green.
Tell them to look where the traps are deep
And the sand flies up in a powdered heap,
And out of the depths loud curses creep
    To the flash of a niblick sheen.

Then if the boss should sigh,
    Or for my presence seek,
Tell him the truth, don't lie;
    Say that my will was weak.
For what is a job to a brassy shot

That whistles away to an untrapped spot,
Or the thrill of a well-cut mashie shot,
  Or the sweep of a burnished cleek?

GRANTLAND RICE (written as the club song of
the St Andrews Golf Club, New York)

A ROUND WITH THE PRO
If any perfection
  Exists on this earth
Immune from correction,
  Unmeet for our mirth,

The despair of the scoffer,
  The doom of the wit,
A professional golfer,
  I fancy, is it.

No faults and no vices
  Are found in this man,
He pulls not nor slices,
  It don't seem he can;

Like an angel from heaven,
  With grief, not with blame,
He points out the seven
  Worst faults in your game.

'You should hold your club *this* way,'
  He tells you, 'not *that*.'
You hold your club his way—
  It hurts you, my hat!

Your hocks and your haunches,
  Your hands and your hips
He assembles and launches
  On unforeseen trips.

He says you should do it
  Like *so* and like *so*;
Your legs become suet,
  Your limbs are as dough.

He tells you to notice
  The way his club wags;
(But how *lovely* his coat is,
  How *large* are his bags!)

[303]

You mark his beginning,
  You watch how he ends,
You observe the ball spinning,
  How high it ascends!

To you the whole riddle
  Is just what he does
When he gets to the middle
  And makes the brute buzz.

He tells you the divot
  You took with your last
Was all due to the pivot—
  Your comment is 'Blast!'

He tells you your shoulders
  Don't sink as they should;
Your intellect moulders,
  Your brains are like wood.

But he pulls his wrists through
  Right under his hands,
His whole body twists through,
  Tremendous he stands.

He stands there and whops them
  Without any fuss;
He scoops not nor tops them
  Because he goes *thus*.

Obsequious batches
  Of dutiful spheres
All day he despatches
  Through Time and the years.

You copy his motions,
  You take it like *this*,
You seize all his notions,
  You strike—and you miss.

You aim with persistence,
  With verve and with flair,
You gaze at the distance—
  The orb is not there.

The hands have been lifted,
  The head remains still,
Your eyes have not shifted—
  No, nor has the pill.

[304]

He points out the errors
　　He told you before,
To add to your terrors
　　He points out two more,

Till, your eyes growing glassy,
　　Your face like a mule's,
You let out with your brassie
　　Regardless of rules.

And the ball goes careering
　　Far into the sky
And is seen disappearing
　　Due south, over Rye.

You stand staring wildly
　　(It's now at Madrid)
And the pro remarks mildly,
　　'You see what you did?

You made every movement
　　I've tried to explain;
That shows great improvement,
　　Now do it again.'

　　　　　　　　E. V. KNOX, *Punch*, 23 October 1929

※

THE GOLFAIYAT OF DUFAR HY-YAM

Myself when young did eagerly frequent
Club-makers' Shops, and heard great Argument
　　Of Grip and Stance and Swing; but evermore
Learned and Bought little I did not repent.

So leave the Cranks to wrangle; and with Me
The Arguments of Theorists let be,
　　And softly by the Nineteenth Hole reclined
Make Game of that which maketh Game of Thee.

For out and in, whichever way we go,
Golf is a pleasant kind of Raree-Show
　　Full of all sorts of unexpected Fun;
(I would not dare to tell Tom Morris so).

The Ball no question makes of Ayes or Noes,
But right or wrong, as strikes the Player, goes;
　　The supercilious Kadi with your clubs
Could tell exactly Why—He knows, He knows!

[305]

Ah, Smith, could thou and I with Fate conspire
To grasp this Game in detail and entire,
   And play it perfectly—would we pursue
Our round with greater Relish and Desire?

Dufar Hy-yam! yet keenness knows no wane;
Behold, the Moon has risen once again;
   How oft, hereafter rising, shall she look
Along these self-same Links for Me—in vain!

When Fate has wound me on her direful Reel,
Should Memory ever to your heart reveal
   A thought of Me when you are passing here—
Press down a DIVOT with a pious Heel.

THOMAS RISK, The Lyre and the Links

THE GOLFER'S DISCONTENT
The evils of the Golfer's state
   Are shadows, not substantial things—
That envious bunkers lie in wait
   For all our cleanest, longest swings;
The pitch that should have won the round
Is caught and killed in heavy ground.

And even if at last we do
   That 80, coveted so long,
A melancholy strain breaks through
   The cadence of our even-song—
A 7 (which was 'an easy 4')
Has 'spoilt our 77 score.'

And thus, with self-deception bland,
   We mourn the fours that should have been,
Forgetting, on the other hand,
   The luck that helped us through the green;
Calmly accepting as our due
The four-hole which we fluked in two.

The drive that barely cleared the sand,
   The brassy-shot which skimmed the wall,
The useful 'kick,' the lucky 'land'—
   We never mention these at all;
The only luck that we admit
Is when misfortune comes of it.

And therefore, in a future state,
  When we shall all putt out in two,
When drives are all hole-high and straight,
  And every yarn we tell is true,
Golf will be wearisome and flat,
When there is naught to grumble at.

Ibid.

BALLADE OF DEAD GOLFERS
Where are the stars that glittered so?
  They wane and flicker, flare and fall;
The champions of long ago
  Are hidden 'neath the common pall.
Great Allan of the feathery ball,
  The sandy hair, the carriage staid,
The twinkling eye—beyond recall:
Their race is run, their round is played.

Piries and Parks and Dunns—we know
  They dared the burn and crossed the wall:
Down swooped the rush of time, and lo!
  They vanished in the midnight squall.
Young Tommy, greatest of them all,
  'Thrice belted knight,' whose towering shade
Out of the past looms vague and tall—
  Their race is run, their round is played.

The slash, the splendour and the glow,
  The swing that held their foes in thrall,
The sheer perfection of the blow
  Must share the master's funeral.
Yet in their day with great or small,
  Holding their own and unafraid,
They raised their best memorial:
  Their race is run, their round is played.

ENVOI

Prince, ne'er discuss—it breeds but gall—
  How haply they would fare with Braid
In some Elysian carnival:
  Their race is run, their round is played.

BERNARD DARWIN

MULLION

My ball is in a bunch of fern,
  A jolly place to be;
An angry man is close astern—
  He waves his club at me.
Well, let him wave—the sky is blue;
Go on, old ball, we are but two—
  We may be down in three,
Or nine—or ten—or twenty-five—
It matters not; to be alive
  Is good enough for me.

How like the happy sheep we pass
  At random through the green,
For ever in the longest grass,
  But never in between!
There is a madness in the air;
There is a damsel over there,
  Her ball is in the brook.
Ah! what a shot—a dream, a dream!
You think it finished in the stream?
  Well, well, we'll go and look.

Who is this hot and hasty man
  That shouteth 'Fore!' and 'Fore!'?
We move as quickly as we can—
  Can any one do more?
Cheer up, sweet sir, enjoy the view;
I'd take a seat if I were you,
  And light your pipe again:
In quiet thought possess your soul,
For John is down a rabbit-hole,
  And I am down a drain.

The ocean is a lovely sight,
  A brig is in the bay.
Was that a slice? You may be right—
  But, goodness, what a day!
Young men and maidens dot the down,
And they are beautiful and brown,
  And just as mad as me.
Sing, men and maids, for I have done
The Tenth—the Tenth!—in twenty-one,
  And John was twenty-three.

Now will I take my newest ball,
  And build a mighty tee,
And waggle once, or not at all,
  And bang it out to sea,
And hire a boat and bring it back,
And give it one terrific whack,
  And hole it out in three,
Or nine—or ten—or twenty-five—
It matters not; to be alive
At Mullion in the summer time,
At Mullion in the silly time,
  Is good enough for me.

A. P. HERBERT, *Mild and Bitter*, 1936

'Seaside Golf' is probably the best-known of the poems about golf and was neatly parodied by Sir Robin Butler at the St Enedoc Centenary Dinner in April 1990.

SEASIDE GOLF

How straight it flew, how long it flew,
  It cleared the rutty track,
And soaring, disappeared from view
  Beyond the bunker's back—
A glorious, sailing, bounding drive
That made me glad I was alive.

And down the fairway, far along
  It glowed a lonely white;
I played an iron sure and strong
  And clipped it out of sight,
And spite of grassy banks between
I knew I'd find it on the green.

And so I did. It lay content
  Two paces from the pin;
A steady putt and then it went
  Oh, most securely in.
The very turf rejoiced to see
The quite unprecedented three.

Ah! seaweed smells from sandy caves
  And thyme and mist in whiffs,
In-coming tide, Atlantic waves
  Slapping the sunny cliffs,

Lark song and sea sounds in the air
And splendour, spendour everywhere.

JOHN BETJEMAN

❖

THE PARODY
How low it flew, how left it flew,
   It hit the dry stone wall
And plunging, disappeared from view
   A shining brand new ball—
I'd hit the damned thing on the head
It made me wish that I were dead.

And up the fairway, steep and long,
   I mourned my gloomy plight;
I played an iron sure and strong,
   A fraction to the right.
I knew that when I reached my ball
I'd find it underneath the wall.

And so I did. I chipped it low
   And thinned it past the pin
And to and fro and to and fro
   I tried to get it in;
Until, intoning oaths obscene,
I holed it out in seventeen.

Ah! seaweed smells from sandy caves
   They really get me down;
In-coming tides, Atlantic waves
I wish that I could drown,
And Sloane Street voices in the air
And black retrievers everywhere.

ROBIN BUTLER, 1990

GOLF AS IT NEVER WAS

❖

There are two classic examples of writing about golf which reveal the authors' absolute ignorance of the subject. In doing so they also underline how little the game was known outside Scotland in the latter part of the nineteenth century.

The first is by W. H. G. Kingston, who became a very well known and successful writer of books for boys. In Ernest Bracebridge, *or* Schoolboy Days, *published in 1860, he attempts to describe in successive chapters how different games are played. With hoops, fishing, fencing, and hare and hounds he is on reasonably safe ground, but when it comes to golf he is lost beyond redemption.*

'You observe, Ellis,' said Bracebridge, 'the great object is to get a ball both hard, light, strong, easily seen, and which will not be the worse for a wetting. All these qualifications are possessed by this little fellow. Why golf has gone out so much in England, I don't know. Two centuries ago it was a fashionable game among the nobility; and we hear of Prince Henry, eldest son of James the First, amusing himself with it. In those days it was called "bandy ball" on account of the bowed or bandy stick with which it was played. We now only apply the term "bandy" to legs. Still further back, in the reign of Edward the Third, the game was played, and known by the Latin name of *Cambuca*. Now are we all ready?'

Macgreggor, who had just come up with his companions, replied that all his party were ready to begin. Each side was accompanied by two boys, carrying a number of other clubs, one of which was of iron, and some were shorter, and some longer, to enable the players to strike the ball out of any hole, or rut, or other place in which it might have got.

'These extra clubs are called putters, and the men who carry them cads or caddies,' Ernest remarked to Ellis. 'This heavy iron club, is, you see, to knock the ball out of a rut, which would very likely cause the fracture of one of our wooden clubs. Now you understand all about the matter. Follow me; I'll tell you what to do when Macgreggor is not near; otherwise though he is playing against us, he will advise us what to do.'

The ball was thrown up and the game began. Macgreggor had the first stroke. He sent the ball a considerable distance towards the nearest hole. Ernest had then to strike his ball. If he struck it very hard it might go beyond the hole, which would have thrown him back; and if he did not send it as far as the ball first struck, Macgreggor's party would have the right to strike twice before his would again strike the ball. Ellis at first thought that there was nothing in the game, but he soon perceived that there was a good deal of science required, and that nothing but constant practical experience could make a person a good player. He however, as Bracebridge was doing, gave his mind entirely to it, and by listening to the remarks made by Macgreggor, he learned the rules and many of the manœuvres golf players are accustomed to practise. He very soon got deeply interested in the game, as did indeed all the party; and perhaps had they been asked at the moment what they considered one of the most delightful things to do all day they would all have pronounced in favour of playing golf.

Golf is a most difficult game to describe. I should liken it, in some respects, to billiards on a grand scale, except that the balls have to be put into holes instead of pockets; that they have to be struck with the side instead of with the end of a club, and that there is no such thing as cannoning.

Bracebridge sent his ball very cleverly a few yards only beyond Macgreggor's, which called forth the latter's warm approval. Then Gregson struck the ball and sent it but a very short distance. Buttar next sent theirs nearly up to the hole, and Bouldon then going on, and being afraid of going beyond the hole, sent it not so far, as Buttar had struck their ball.

'Two, two,' shouted Bracebridge. 'Now, Knowles, hit very gingerly and let me see if I cannot send our ball in.'

Knowles rolled the ball within a few feet of the hole, and Ernest, who in consequence of Bouldon's miss, was now allowed to strike, guided by his correct and well-practised eye, sent it clean into the hole, to the great delight of Macgreggor, who was pleased at having so apt a pupil. Bracebridge now took the ball out of the hole, and struck it on. Macgreggor, however, was not long in catching him up, but Tom Bouldon was a great drawback to Macgreggor, he had not calmness to play the game well. He was continually missing the ball, or sending it beyond the hole, while Macgreggor and Bracebridge, and Ellis especially, always considered how far it was necessary to send it, and took their measures accordingly.

Few games show the character of a person more than does that of golf, although all, more or less, afford some index to those who are attentively looking on. A boy, when playing, should endeavour to keep a watch over himself as much as on all other occasions, and he should especially endeavour to practise the very important duty of restraining his temper. Boys are too apt to fancy that they may say and do what they like, and often they abuse each other, and make use of language of which, it is to be hoped, they would be ashamed when out of the playground.

W. H. G. KINGSTON, Ernest Bracebridge, or Schoolboy Days, 1860

It may not be surprising in view of the above that the reactions of the other boys who were watching the game were not favourable:

While the game was going on and drawing near to its conclusion, Bracebridge being ahead, a number of boys came to see what was going forward. From their remarks, there was not much chance of the game becoming popular. There was not enough activity in it to please them. It was not to be compared for a moment with cricket or rackets or football or even hockey.   Ibid.

The second example is from the Philadelphia Times, 1889. *Here again the account is so wayward that one is tempted to assume that the author was the victim of a practical joke; if that was indeed the case, it was certainly entirely successful, since there is no doubt that the piece was written in good faith.*

Up to this time golf has made little way in the United States. It is occasionally played in Canada, although even there it has not assumed the importance of a regular department of sports. It is a game that demands at once the utmost physical development upon the part of the player as well as a considerable amount of skill, and it arouses the interest only of those who go into sports for the love of action. It is far from being a 'dude' game. No man should attempt to play golf who has not good legs to run with and good arms to throw with, as well as a modicum of brain power to direct his

[313]

play. It is also, by the nature of the game itself, a most aristocratic exercise, for no man can play at golf who has not a servant at command to assist him. It is probable that no sport exists in the world to-day or ever did exist in which the services of a paid assistant are an essential as in this national game of Scotland. The truth is that the servant is as essential to the success of the game as the player himself.

To play golf properly there is needed a very large expanse of uncultivated soil, which is not too much broken up by hills. A few knolls and gulleys more or less assist to make the game more interesting. In Scotland it is played generally upon the east coast, where the links are most extensive. Having selected a field, the first thing necessary is to dig a small hole, perhaps one foot or two feet deep and about four inches in diameter. Beginning with this hole a circle is devised that includes substantially the whole of the links. About once in 500 yards of this circle another hole is dug. If the grounds selected cannot include so large a circle as this, the holes may be put at as short a distance as 100 yards from each other; but the best game is played when the field is large enough to include holes at a distance of 500 yards apart. The game then may be played by two or four persons. If by four, two of them must be upon the same side.

There are eleven implements of the game, the most important of which is the ball. This is made of gutta percha and is painted white. It weighs about two ounces and is just small enough to fit comfortably into the holes dug in the ground. Still it should not be so large that it cannot be taken out with ease. The other ten implements are the tools of the players. Their names are as follows: the playing club, long spoon, mud spoon, short spoon, baffing spoon, driving putter, putter, sand iron, club and track iron. Each of these is about four feet long, the entire length of which in general consists of a wooden handle. The head is spliced on, and may be either metal or wood. The handle, as a rule, is made of hickory covered with leather.

At the beginning of play each player places his ball at the edge of a hole which has been designated as a starting point. When the word has been given to start he bats his ball as accurately as possible towards the next hole which is either 100 or 500 yards distant. As soon as it is started in the air he runs forward in the direction which the ball has taken, and his servant, who is called a 'caddy' runs after him with all the other nine tools in his arms. If the player is expert or lucky he bats the ball so that it falls within a few feet or inches even of the next hole in the circle. His purpose is to put the ball in the next hole, spoon it out and drive it forward to the next further one before his opponent can accomplish the same end. The province of the 'caddy' in the game is to follow his master as closely as possible, generally at a dead run, and be ready to hand him whichever implement of the game the master calls for, as the play may demand. For instance the ball may fall in such a way that it is lodged an inch or two above the ground, having fallen in thick grass. The player, rushing up to it, calls on his 'caddy' for a baffing spoon, and having received it from the hands of his servant he bats the ball with the spoon in the direction of the hole. An inviolable rule of the game is that no player shall touch the ball from one limit of the circle to the other with his hands. All play must be done with the tools.

In this the caddy really gets about as much exercise out of the sport as his master, and he must be so familiar with the tools of the game that he can hand out the right implement at any moment when it is called for. If the player has succeeded in throwing or pushing his ball into a hole, his opponent must wait until he has succeeded in spooning it out before he begins to play. Obedience to this rule obviates any dispute as to the order in which a man's points are to be made. For if one player has his ball in a hole and his opponent has his within an inch or two of it, he must wait before he plays until the first player has gotten his ball clear of it and thrown it towards the next hole. Following this general plan the players go entirely about the circle, and in a large field this may involve a run of several miles. If the ball is thrown beyond the hole, it must be returned to it and carefully spooned out again. The aim of the sport is not necessarily to complete the circle as quickly as possible. There are no codified rules according to which the game is played. As a general custom the players make the entire circuit of the circle and the one who gets his ball in the hole at which they began first wins the game. Nevertheless it is sometimes agreed that the game shall be won by him who makes the largest number of holes within a given number of minutes, say twenty or thirty. In either case the principle of the game remains the same; and if partners are playing, it simply means that if A strikes a ball and B is his partner, B must run forward and make the next play, and *A* must run after him and make the next, and so on, while *D* and *C*, who are on the other side, are doing the same thing. In this partnership game there is actually more exercise to the players than in the single game and the servants or 'caddies' are equally busy.

Spectators sometimes view games of golf, but as a rule they stand far off, for the nature of the implements employed is such that a ball may be driven in a very contrary direction to that which the player wishes, and therefore may fall among the spectators and cause some temporary discomfort. Moreover it would require considerable activity upon the part of the spectators to watch the play in golf, for they would have to run around and see how every hole was gained, from one end of the game to the other. There may be as many as thirty spectators at one game, but seldom more, and a good game is frequently played without any at all.

The principal qualifications for the game are steady nerve and eye and good judgment and force with an added ability to avoid knolls and sand-pits which, in the technical terms of the Scotch game, are called hazards.

It is not a game which would induce men of elegance to compete in, but those who have strong wind and good muscle may find it a splendid exercise for their abilities, and plenty of chance to emulate each other in skill and physical endeavour.

PHILADELPHIA TIMES, 24 February 1889

SOURCE
ACKNOWLEDGEMENTS

❖

Julian Barnes, from *Staring at the Sun* (Cape, 1986). Reprinted by permission.

Sir John Betjeman, from *Collected Poems*. Reprinted by permission of John Murray (Publishers) Ltd.

Tom Bridges, from *Alarms and Excursions* (Longman, 1938).

Valentine E. C. Browne (Lord Castlerosse), from *Valentine's Days* (Methuen, 1934).

Patrick Campbell, from *How to Become a Scratch Golfer* (Blond, 1963). Copyright the Estate of Patrick Campbell.

Alastair Cooke, from *The Bedside Book of Golf*, ed. Donald Steel (B. T. Batsford, 1965). Copyright Alastair Cooke 1965.

Bernard Darwin, from *Golf between Two Wars* (Chatto & Windus, 1944).

David Davies, from the *Guardian*, 8 April 1992. © *Guardian* 1992.

Patric Dickinson, from *The Good Minute* (Gollancz, 1965). © Patric Dickinson 1965; from *A Round of Golf Courses* (A & C Black, 1951). Copyright Patric Dickinson 1965. Used with permission.

Ian Fleming, from *Goldfinger* (Cape, 1959). © Glidrose Publications Ltd. 1959. Reprinted by permission of Glidrose Publications Ltd.

George MacDonald Fraser, from *McAuslan in the Rough* (Barrie & Jenkins, 1974).

A. P. Herbert, from *The Rules of the Game*, in *The Punch Book of Golf*; from *The Man About Town* (Heinemann, 1923); 'Mullion' from *Mild and Bitter* (1936). Reprinted by permission of A. P. Watt Ltd.

Dan Jenkins, from *The Dogged Victims of Inexorable Fate* (Little Brown & Co., 1970). © Dan Jenkins 1970.

William Oscar Johnson & Nancy Williamson, from *Babe*, Part 3, *Sports Illustrated*, 20 October 1975. © William Oscar Johnson & Nancy Williamson 1975.

Bobby Jones, from *Golf is my Game* (Chatto & Windus, 1959); from *Down the Fairway* (Allen & Unwin, 1927); from *The Greatest of Golfers*, in the *American Golfer*, August 1930.

E. V. Knox, from *The Punch Book of Golf*. Used with permission.

SOURCE ACKNOWLEDGEMENTS

Henry Leach, from *The Spirit of the Links* (Methuen, 1907).

Henry Longhurst, from article in the *Sunday Times*; from *Round in Sixty-Eight* (Werner Lawrie, 1953).

A. G. MacDonell, from *England, their England* (Macmillan Ltd., 1933).

A. A. Milne, from *Not that it Matters* (1919). Copyright the Estate of A. A. Milne.

E. Phillips Oppenheim, from *The Pool of Memory* (Hodder & Stoughton, 1941).

Francis Ouimet, from *A Game of Golf* (Hutchinson, 1933). Reprinted by permission of Random House UK Ltd.

Walker Percy, from *The Second Coming* (Secker & Warburg, 1981). © Walker Percy 1981. Used with permission.

Chris Plumbridge, from an article in the *Sunday Telegraph*, 1992. © Chris Plumbridge.

Stephen Potter, from *The Complete Golf Gamesmanship* (Heinemann, 1968). Copyright © 1968 Stephen Potter.

Lorne Rubinstein, from *Links* (Prima Publishing, 1991). © 1991 Lorne Rubinstein. Reprinted by permission of the publisher.

Sapper (H. H. McNeill), from *Uncle James's Golf Match* from *Out of the Blue* (Hodder & Stoughton). Used with permission.

Gene Sarazen, from *30 Years of Championship Golf* (Prentice-Hall, 1950). Copyright Gene Sarazen 1950.

Siegfried Sassoon, from *The Weald of Youth*. Reprinted by permission of G. T. Sassoon.

John Uplike, from *Rabbit is Rich* (Andre Deutsch, 1982); from *Golf Dreams*, first published in the *New Yorker*; from *The Golf Course Owner*, first published in the *Ontario Review*. Reprinted by permission.

Pat Ward-Thomas, from *The Long Green Fairway* (Hodder & Stoughton, 1966); from *Masters of Golf* (Heinemann, 1961).

Joyce Wethered, from *Golfing Memories and Methods* (Hutchinson, 1934).

Herbert Warren Wind, from *Following Through* (Ticknor & Fields, 1985). © Herbert Warren Wind 1985. Reprinted by permission.

P. G. Wodehouse, from *Chester Forgets Himself*, in *Omnibus of Golf Stories* (1973). Reprinted by permission of A. P. Watt Ltd.

Any errors or omissions are entirely unintentional. If notified the publisher will be pleased to make any necessary amendments at the earliest opportunity.

INDEX OF NAMES

❖